DIGITAL PEOPLE

DIGITAL PEOPLE

FROM BIONIC HUMANS TO ANDROIDS

by
Sidney Perkowitz

Joseph Henry Press
Washington, D.C.

Joseph Henry Press • 500 Fifth Street, N.W. • Washington, D.C. 20001

The Joseph Henry Press, an imprint of the National Academies Press, was created with the goal of making books on science, technology, and health more widely available to professionals and the public. Joseph Henry was one of the founders of the National Academy of Sciences and a leader in early American science.

Library of Congress Cataloging-in-Publication Data

Perkowitz, S.
Digital people : from bionic humans to androids / Sidney Perkowitz.
p. ; cm.
Includes bibliographical references and index.
ISBN 0-309-08987-5 (hardback)
1. Robotics. 2. Artificial intelligence. I. Title.
TJ211.P44 2004
629.8'92—dc22

2004000049

Printed in the United States of America.

To Sandy and Mike, with love, . . . again

Contents

1 Introduction: Androids All Around Us 1

PART I: ARTIFICIAL BEINGS: MEANING AND HISTORY

2 The Virtual History of Artificial Beings 17

3 The Real History of Artificial Beings 51

4 We Have Always Been Bionic 85

PART II: HOW FAR ALONG ARE WE?

5 Mind-Body Problems 105

6 Limbs, Movement, and Expression 123

7 The Five Senses, and Beyond 147

8 Thinking, Emotion, and Self-Awareness 173

9 Frankenstein's Creature or Commander Data? 199

Suggested Reading 220

Filmography 225

Acknowledgments 227

Index 229

1

Introduction: Androids All Around Us

You might not know it, but among us there exist artificial beings that are lifelike enough to give you goose bumps. If you had visited robot developer Rodney Brooks at MIT in the late 1990s, you would have met his Cog (short for "cognitive") robot. Shaped vaguely like a human head and torso, and built more or less to human scale, Cog still looked alien and machinelike because it was made of girders and electronic components. Instead of eyes, video cameras located in its head fed visual information to its computer brain.

But when I saw Cog's intricate body language, I forgot its machine appearance. Although it did not look like a person, it acted like one. Those sensors and computers, motors and metal supports kept its "eyes" in continual motion, scanning the scene for interesting events—just as our own brains and eyes do at an unconscious level. And when the door opened and a student walked in, Cog did what you or I would; it stopped scanning and turned toward the visitor. As Cog brought its gaze and (apparently) its full attention to bear on her, the action was so eerily human that it gave me a moment of hair-raising, gut-level understanding, for in that instant, Cog seemed fully alive and conscious.

A year or two later, in that same laboratory, you would have met another robot, Kismet, created by Cynthia Breazeal, Brooks's graduate student at the time, now an MIT professor and well-known robotics researcher in her own right. Where Cog is intimidating, Kismet has a face out of a children's storybook, clownlike and cartoonish with exaggerated features—huge blue eyes, bright red lips, and prominent, highly mobile ears. Approach Kismet and engage its attention by waving a toy or talking, and it responds in a tiny voice, moving its head, and adjusting its face to smile, or to look sad, angry, or fearful. When Kismet was young, Breazeal brought in adults and children to interact with it. Today she says, "Kismet became a personality to them, to the point where people still ask me 'How's Kismet?' They refer to Kismet as a creature rather than this thing in the lab."

Human reactions to Cog and Kismet offer an important lesson: regardless of what is going on inside an artificial being—and the debate over what might constitute "machine intelligence" and "machine consciousness" is a deep and continuing one—the merest hint of humanlike action or appearance deeply engages us. Cog generates a surprising sense of life simply through its reactions to its environment. Kismet goes further; it reacts to people with, for example, facial expressions that humans sense in a direct and natural way.

Other artificial creatures add vigorous body movements or other levels of interaction. At the Honda research laboratories in Japan, a child-size robot, humanoid in outline, walks, balances on one foot, and nimbly climbs stairs without a hitch. At MIT, Carnegie Mellon University, and the Palo Alto Research Center, artificial creatures roll, slither, crawl, stride, and hop across the floor, or configure and reconfigure their bodies so as to locomote in the most efficient way. At the ROBODEX 2003 exposition in Yokohama, Japan, robots answered questions, reacted to human body language, sang, danced, and played soccer. At Walt Disney theme parks in the United States and Europe, and in countless Hollywood movies, entertainment androids convincingly simulate people, animals, and imaginary beings.

Most remarkably, artificial creatures are beginning to generate a kind of emotional lifelikeness because they create warm feelings in people, as you can easily see without visiting any robotics laboratories. Just spend a few moments with any recent robotic toy such as the Sony Corporation's AIBO dog (**A**rtificial **i**ntelligence + ro**bo**t) that went on sale in 1999, or the I-Cybie robot dog, made by Tiger Electronics and Silverlit Toys. These two could never be taken as natural creatures because both are only plastic caricatures of a dog, but like Cog, they need not work very hard to elicit human reactions. If the creature interacts with the world, has some capacity to change its behavior as it gains experience—that is, if it can learn—and displays natural-seeming behavior, it can project a well-nigh irresistible impression of life.

As you watch I-Cybie cock its head toward you when you call its name, or AIBO perform a trick at your voice command, it's easy to feel *something* toward the mechanism: amazement that it listens to you or a small rush of affection. And if the synthetic being looks like a human rather than an animal, like Kismet's face or the toy robot infant called My Real Baby released in 2000, its emotional power is far more intense.

On the face of it, it might seem unreasonable to have feelings toward "a creature that really doesn't know you're there," as sociologist Sherry Turkle of MIT puts it, yet it happens all the time. Little girls have always loved their dolls, no matter how crude, and children and adults bond to objects and machines not in the least cute or petlike. We become attached to bicycles, boats, and computers, giving them names, endowing them with personalities, and projecting human or animal dimensions onto their actions. We swear at a "stubborn" or "cranky" lawnmower that won't start, or affectionately caress a sleek car as we would a superb racehorse.

Artificial beings, however, are not limited to fully manufactured creatures of plastic and metal. We ourselves are partly artificial or "bionic"—that is, people with synthetic parts—to a surprising extent: 8 to 10 percent of the U.S. population, approximately 25 million people, and becoming more so as our population ages. Our bionic additions include functional prosthetic devices and implants, such as artificial limbs, replacement knees and hips, and vascular stents (tiny gridlike metallic inserts that aid the flow of blood in blocked arteries, in themselves a multibillion dollar industry). There are also cosmetic or vanity bionic additions that replace what accident or nature took away, or genetic inheritance never gave, from false hair and teeth to artificial eyes and breast augmentations (more than 200,000 of which were implanted in 2001 alone).

Cosmetic additions like these might be used in a project to make an artificial creature look human, but bionic alterations beyond the merely cosmetic are significant steps toward building a whole creature. Where once we had only crude prosthetic devices such as a glass sphere in place of an eye, or an iron hook for a hand, now we are developing functional body parts that are increasingly indistinguishable from the real thing, some with neural connections—not only limbs, but replacements for lost or diseased vision, hearing, and other capacities and organs. Most startling of all, we now look beyond the physical to "bionic brains" that is; "artificial" means to alter or augment mental capacity and emotional states, from implanted drug delivery systems to computer chips connected directly to the neural network.

The combination of human with artificial components lies at one end of a spectrum of artificiality, depending on how much of the being is made of natural or living parts, or is meant to look natural, and on how self-directed the creature is—from automaton, to robot, to android, to cybernetic organism or cyborg, to bionic human. An automaton is a machine that appears to move spontaneously, although actually it moves "under conditions fixed for it, not by it" according to one definition. A robot is an autonomous or semiautonomous machine made to function like a living entity (here, "machine" includes mechanical, inorganic, or organic but nonliving moving or static parts, and electronic, computational, and sensory components). It can be humanoid, although not necessarily so; most contemporary robots take nonhuman shapes that are useful for their particular applications. An android is similarly entirely artificial but has been made to look human (the word comes from Greek roots meaning "manlike."). *Star Trek: The Next Generation*'s Commander Data is a popular example of an android.

A cyborg (cybernetic organism) and a bionic human (from "biological" and "electronic") are different from the previous three categories, in that both involve a combination of machine and living parts. In my usage, a cyborg has a machine portion that might dominate the natural part in mass and bulk but is under the mental direction of the natural part—essentially, a brain in a box. A bionic human, on the other hand, is mostly natural with a relatively small portion given over to implants or replacement parts such as a heart pacemaker or an artificial limb.

The categories from automaton to bionic human include mobile and responsive robots, amusingly lifelike toys, entertainment androids, humans with mechanical and electronic implants, and others. All are part of a technology that is beginning to realize an extraordinary achievement: the creation of partly or fully artificial beings. Although these possibilities draw on the ultimate in twenty-first-century science, they are not new in the collective human imagination; artificial beings have intrigued, terrified, and exalted us for millennia.

The reasons for this long-standing interest are basic to human nature, although it is not easy to say which of the reasons dominate. Least noble, perhaps, but understandable, is the desire to ease our lives by creating workers to till our fields, operate our factories, and prepare our meals, tirelessly and without complaint. As long ago as the fourth century BCE no less a thinker than Aristotle saw the potential for automated machinery to reduce labor, and even its potential for disrupting the job market:

> If every instrument could accomplish its own work, obeying or anticipating the will of others . . . if the shuttle could weave, and the pick touch the lyre, without a hand to guide them, chief workmen would not need servants, nor masters slaves.

Aristotle's idea was perhaps first realized in eighteenth-century France in an innovative and efficient automated loom for silk weaving. The silk workers immediately understood that the device meant the loss of their livelihoods and objected to its adoption. This is one example of the contradictory quality typical of many aspects of artificial beings (and indeed of all technology): the good that they might bring is counteracted by undesirable side effects that might ensue.

At another level, perhaps nobler, perhaps only a matter of enlightened self-interest, is the desire to transcend our limitations: We imagine creating beings that go beyond humanity's natural physical and mental endowments. A related motivation is the desire to bionically repair ill or damaged bodies and minds or to enhance them for better performance, improved health, and longer lifespan—or, returning to the cosmetic theme, for greater beauty. These, too, are old ideas. Indian mythological writings 4,000 years old or older tell of a warrior queen who went into battle with a prosthetic iron leg, and in Norse mythology, Sif, wife of the god Thor, had dwarves make golden hair for her.

The desire to make ourselves healthier and more beautiful is rooted in our strongest motivation to consider artificial beings: fear of death. It is extreme fantasy, perhaps, to think that artificial creations might allow individuals or the entire race to foil nature and achieve immortality; but it is no fantasy to say that as we develop such beings, we begin scientifically exploring the incomprehensible gap between the living and dead, the animate and inanimate.

Inevitably, even cautious forays into this territory carry a scent of hubris, in the belief that we can outdo evolutionary forces or perhaps God Himself. As the science fiction author Stanislaw Lem has written: "The concept of an artificially created man is blasphemy in our cultural sphere. Such a creation [is] a caricature, an attempt by humans to become equal to God." From the viewpoint of traditional religion, he adds, this blasphemy could succeed only if humanity were to collaborate with the forces of evil; that is why an air of the uncanny surrounds these creatures.

For those who find this eeriness unsettling or the blasphemy unforgivable, other motives for making artificial creatures might prove compelling. Beyond physical improvements, perhaps we can create beings or states of consciousness that avoid our moral and spiritual failings, thereby guiding us toward becoming better humans. And from the scientific viewpoint, what could be more important than to understand the origins and processes of life? In this spirit, research on artificial beings is a way to express our sense of wonder about life and our place among the living, and to better understand both. And surely the potential medical benefits to humanity cannot be dismissed.

Nevertheless, when considering the creation of artificial beings, we must also consider the ambiguities and dark notes inherent in the quest. Whenever such creatures seem to cross the boundary between the living and the dead, the result is awesomely frightening, as shown in a tale from Roman times told in Gerard Walter's biography of Julius Caesar. Supposedly, at the assassinated Caesar's funeral, the crowd suddenly experienced "a vision of horror [and] brutality" when

> From the bier Caesar arose and began to turn around slowly, exposing to their terrified gaze his dreadfully livid face and his twenty-three wounds still bleeding. It was a wax model which [Marc] Antony had ordered in the greatest secrecy and which automatically moved by means of a special mechanism hidden behind the bed.

In a similar vein, in Mary Shelley's *Frankenstein* (1818), when Victor Frankenstein sees the first stirrings of the being he has created from dead body parts, he is shocked and horrified and spontaneously rejects his creation.

These visceral reactions represent the deepest fears that artificial beings might engender. But not every such creature represents a direct challenge to death or to God's law—and if that challenge is absent, so is supernatural fear. When Sigmund Freud addressed this sense of dread in his essay "The 'Uncanny,'" he did not relate it to religious guilt about blasphemy, but to knowledge of our own mortality. We feel uncanny, he says, when a deep emotion that has been repressed is made to recur. Our feelings about death are like that. Children, Freud notes, unambiguously want their inanimate dolls to come to life. Children, however—at least very young ones—have no knowledge of death. Adults do, and as Freud says, because of the "strength of our initial emotional reaction to death and the insufficiency of our scientific knowledge about it . . . almost all of us still think as savages do on this topic." And so a special eeriness arises in the presence of a dead body or when we wonder whether something seemingly dead, such as an automaton, is actually alive.

If Freud is correct, then research on artificial beings can only reduce the sense of uncanniness as it explores the borderline between the living and the dead. In any case, the technological creation of beings from inert metal, plastic, and silicon is a different matter from animating the dead. Perhaps that explains why technologists seem unconcerned about blasphemy as they try to create synthetic beings. One young researcher in the field recently summed it up when she said, "I thought it would be neat to design something that reproduces what God can do." Call her attitude what you will—hubris, or a healthy pride in science—the scientists and engineers spearheading the creation of artificial beings and bionic people are responding to the magnetism of the technological imperative, the pull of a scientific problem as challenging as any imaginable.

Fascinating scientific puzzle though it is, the creation of artificial beings is also expected to meet important needs for society and individuals. Industrial robots are already widely used in factories and on assembly lines. Robots for hazardous duty, from dealing with terrorist threats to exploring hostile environments, including distant planets, are in place or on the drawing boards. Such duty could include military postings because there is a long-standing interest in self-guided battlefield mechanisms that reduce the exposure of human soldiers, and in artificially enhanced soldiers with increased combat effectiveness. (For this reason, the Department of Defense, largely through its research arm—the Defense Advanced Research Projects Agency (DARPA)—is the main U.S. funding source for research in artificial creatures.) Artificial creatures can also be used in less hostile environments: homes, classrooms, and hospitals and rest homes, serving as all-purpose household servants, helping to teach, and caring for the ill or elderly.

Among these possibilities, the connection between artificial creatures and human implants might be the most important because it promises enormous medical benefits. This connection might be the single greatest motivation to develop artificial beings. Yet regardless of their potential good uses, and apart from any issues of blasphemy, we have concerns about robots and androids. One fear is that the limitations we think to design out of our creations, from cosmetic deficiencies to the existential realities of illness and death, are essential human attributes, and that to abandon them is somehow to abandon our humanity. Something in us, it seems, fears perfection, and artificial beings threaten us with an unwelcome perfection, expressed as rigid unfeeling precision.

There is another menace first conveyed nearly 200 years ago in *Frankenstein,* and now more compelling than ever: the fear that technology will grow out of control and diminish humanity for all of us. That concern is hardly limited to artificial creatures. It appears in many arenas—the loss of privacy associated with new forms of surveillance and data manipulation; the depersonalization of human relationships; the incidence of human-made ecological disaster; the growing gap between the world's technological "haves" and "have-nots." It is especially and deeply unsettling, however, to contemplate the literal displacement of humanity by beings made in the human image, only better.

Although *Frankenstein* is the most famous story touching on many of these matters, it is not the only one. The depth of our reactions is shown in a whole imaginative narrative of artificial beings—a millennia-old fantasy or "virtual" history, in which these creatures are the focus of a panoply of emotions, hopes, and concerns. In one thread of the virtual history, humans develop strong feelings for inanimate or artificial beings, as in the Greek myth of Pygmalion, who yearns for his statue of a beautiful woman to come alive. That thread also appears in E.T.A. Hoffman's nineteenth-century story "The Sandman," where a young man falls in love with a clockwork automaton, and in the classic 1982 science-fiction film *Blade Runner,* where a special agent dedicated to the destruction of androids falls in love with one of them. In another thread in the virtual history, artificial beings yearn to become human or accepted as human, for example the "monster" in *Frankenstein,* the puppet Pinocchio, Commander Data in *Star Trek,* and the little boy android in the 2001 film *A. I.: Artificial Intelligence.*

In yet other stories, robots display intelligence and ethical standards that make them trusted guides to a better future for humanity, as in Isaac Asimov's book *I, Robot,* but in a contrary thread, other equally able robots and androids slaughter people, as in Karel Capek's play

R.U.R. and the recent *Terminator* films. And even if artificial beings do not wish to wipe us from the earth, their superiority might still destroy us by stifling human creativity and independence, as in Jack Williamson's story "With Folded Hands."

No current artificial creatures can carry out these scenarios, nor are there yet bionic humans or cyborgs who are the physical or mental superiors of natural people. The abilities of robots and androids are still limited. If they behave intelligently, they do so only in specialized areas, or at a childlike rather than an adult level; though they might be mobile, they cannot yet independently navigate any arbitrary room or street; they are not conscious and self-aware, and hence are not moral beings as we understand morality; they are not emotional, and although they might elicit affection or an appreciation of cuteness as a living pet does, they evoke no deeper feelings.

They cannot pass for human in either appearance or behavior, at least not at the behavioral level proposed by the British mathematician, Alan Turing, in 1950. In what is now universally known as the Turing test, he proposed a purely verbal criterion for defining a "thinking machine" as intelligent. Imagine, he said, that a human observer can communicate with either the machine or another human without seeing either (for instance, via keyboard and printer), and can ask either any question. If after a reasonable time the observer cannot identify which of the two is the computer, the machine should be considered intelligent.

Some researchers now think the Turing test is not a definitive measure of machine intelligence. Yet it still carries weight, and now, for the first time in history, the means might be at hand to make beings that pass that test and others. Advances in a host of areas—digital electronics and computational technology, artificial intelligence (AI), nanotechnology, molecular biology, and materials science, among others—enable the creation of beings that act and look human. At corporations and academic institutions around the world, in government installations and on industrial assembly lines, artificial versions of every quality that would make a synthetic being seem alive or be alive—intelligent self-direction, mobility, sensory capability, natural appearance and behavior, emotional capacity, perhaps even consciousness—are operational or under serious consideration.

Not everyone engaged in these efforts is a robotics engineer or computer scientist. Researchers in other fields are working to help ill and injured people: Some of the most exciting efforts are in biomedical research laboratories, in hospitals and clinical settings, where physicians and engineers are developing artificial parts, such as retinal implants for the blind, that might eventually enhance human physical and mental functions. The medical applications and the engineering technologies enhance each other, and as they grow together, the potential for therapeutic uses brings significant motivation and a clear moral purpose to the science of artificial beings.

There is, however, considerable debate about the possibility of achieving the centerpiece of a complete artificial being, artificial intelligence arising from a humanly constructed brain that functions like a natural human one. Could such a creation operate intelligently in the real world? Could it be truly self-directed? And could it be consciously aware of its own internal state, as we are?

These deep questions might never be entirely settled. We hardly know ourselves if we are creatures of free will, and consciousness remains a complex phenomenon, remarkably resistant to scientific definition and analysis. One attraction of the study of artificial creatures is the light it focuses on us: To create artificial minds and bodies, we must first better understand ourselves.

While consciousness in a robot is intriguing to discuss, many researchers believe it is not a prerequisite for an effective artificial being. In his *Behavior-Based Robotics*, roboticist Ronald Arkin of the Georgia Institute of Technology argues that "consciousness may be overrated," and notes that "most roboticists are more than happy to leave these debates on consciousness to those with more philosophical leanings." For many applications, it is enough that the being *seems* alive or *seems* human, and irrelevant whether it *feels* so. Even our early explorations of artificial beings show us that the goal of seeming alive and human might be less challenging than we might expect because—for reasons only partly apparent—we tend to eagerly embrace artificial beings. As in the common reaction to Kismet or the robotic dogs, it takes only a few cues for us to meet creatures halfway, filling in gaps in their apparent naturalness from the well of our own humanity. In a way, an artificial being exists most fully not in itself, but in the psychic space that lies between us and it.

And yet . . . there is the dream and the breathtaking possibility that humanity can actually develop the technology to create qualitatively new kinds of beings. These might take the form of fully artificial, yet fully living, intelligent, and conscious creatures—perhaps humanlike, perhaps not. Or they might take the form of a race of "new humans" that is; bionic or cyborgian people who have been enormously augmented and extended physically, mentally, and emotionally.

New humans could also arise from a different thread in modern technology. Purely biological methods such as cloning, genetic engineering, and stem-cell research offer another way to enhance human well-being and change our very nature. While astonishing progress has been made in these areas, we have yet to see definitive, broad-scale results. Moreover, a program for changing humans at the genetic level has ethical and religious implications that trouble many people, and the consequences of human-induced changes propagating in our gene pool trouble many scientists. The creation of fully or partly artificial beings has its own set of moral issues; these, however, might ultimately prove more acceptable to society than those arising from genetic manipulation.

At its furthest reach, and as a great hope for the technology of artificial beings, we might be able to create a companion race—self-aware and self-sufficient, perhaps like us in some ways but different in others, with its own view of the universe and new ways to think about it. Fascination with the notion of communicating with another race of beings has been a main incentive in the search for intelligent life elsewhere in the universe—a hope that engages many people, as witness the great interest in the 1996 announcement that traces of ancient life were found on Mars. But that announcement was mistaken, and although the search continues (for instance, with the 2004 landing of two NASA robotic planetary explorers on Mars), we have yet to find evidence of alien beings anywhere that our spacecraft and telescopes can reach. Perhaps we never will, so the creation right here on Earth of a race that complements humanity has special appeal.

No matter what emerges from controversies about robotic consciousness or the morality of making artificial beings, no matter what approach to artificial intelligence proves effective, one thing is clear: Without digital electronics and digital computation, we could not begin to consider artificial intelligence and artificial sensory apparatus, the physical control of synthetic bodies, and the construction of interfaces between living and nonliving systems. Although the history of artificial beings has presented many ways to create them, animate them, and give them intelligence, now we are truly entering an era of digital people.

Part I

Artificial Beings: Meaning and History

In 1950, Alan Turing opened his seminal paper that defined the Turing test with the provocative sentence "I propose to consider the question 'Can machines think?' " More than a half century later, I propose a new question: Can machines live? It is a fantastic question and its answers can come only in parts—some connected to technological realities and some indeed connected to fantasy, the virtual history of imaginary artificial creatures, where we seek our first set of answers.

2

The Virtual History of Artificial Beings

The intensity of our interest in artificial beings is due to the compelling meanings we attach to them. Technology has yet to give us perfect replacement body parts or full-fledged androids, but millennia ago, the cultural repositories of our dreams and self-images—legend, myth, and eventually written literature—presented a rich account, a virtual history, of imaginary artificial beings. Later, the tales were told in new formats, like film and television. In all ages and media, the stories we built around these would-be creatures express our desires and fears, define the expectations we place on these beings, and create the vocabulary we use to describe them.

In a way, these fantasy versions are now building themselves into reality because many of today's creators of artificial beings owe their passion to childhood encounters with robots and androids in science fiction and fantasy. The connection flows the other way too, because science influences works of the imagination. Eighteenth-century studies of electricity played into Mary Shelley's *Frankenstein;* today's technology inspires the artificial beings depicted in the entertainment media, from the robot R2D2 in *Star Wars* to the child android in *A. I.: Artificial Intelligence.*

The most powerfully symbolic of these virtual life forms is the creature Victor Frankenstein created. Others stand out as well, such as the robots in Karel Capek's 1921 play *R.U.R.*, Fritz Lang's 1927 film *Metropolis,* and Isaac Asimov's 1950 book *I, Robot.* More recently, there are the androids in the 1982 film *Blade Runner*, bionic humans in the television series *The Six Million Dollar Man* (1973–1978), and a cyborg in the 1987 film *RoboCop.*

These tales illuminate every aspect of our complex thoughts and feelings, many of them contradictory, about artificial life. There is the visceral dread that envelops us as Frankenstein's creature stirs into life, that deep fear of stepping across the boundary between the living and nonliving. Yet we also feel compassion for the creature, as we do for the cyborg in the film *RoboCop,* who retains painful human emotions. At the same time, we admire RoboCop's moral strength and reliability. Other creatures, from the manipulative female robot in *Metropolis* to the murderous androids in the *Terminator* films, act evilly. Some carry no special moral stance, but bring us beauty, like the cyborg dancer Deirdre in the story "No Woman Born" by C.L. Moore. Some are loved, and perhaps return love, like the android Rachael in *Blade Runner.* Examining these imaginary beings helps us understand our motives for making them, and predicts the attitudes we bring to their actual creation.

CREATURES OF BRONZE AND CLAY

Fears and dreams of artificial beings go far back, at least to the legend of Pygmalion the sculptor, an ancient Greek vision of inanimate matter coming alive. Pygmalion made an ivory statue of a beautiful woman and came to love it. One day he returned from a festival in honor of the goddess Aphrodite, kissed the statue, and found to his delight that it turned into a warm and living woman, whom he soon married.

In the myth, Aphrodite brings the statue to life, in response to Pygmalion's yearnings. Today, we expect technology rather than a god to intervene. The Greeks, too, recognized technology (the very word is Greek in origin) in another myth about a self-acting being made of metal. The story involved the deity who could be called the Greek god of technology—Hephaestus, who was in charge of fire and the forging of metal, and whom the Romans called Vulcan.

According to Homer, Hephaestus was the son of Hera and Zeus. Others say Hera alone conceived and bore Hephaestus, with no intervention from Zeus or any other partner, to spite Zeus after he had fathered Athena alone. Whatever his origin, Hephaestus was the limping god, born with a lame leg and a clubfoot, who walked with a crutch. As artificer to the gods, he made marvelous contrivances such as Achilles' shield and Apollo's chariot. This legend of a handicapped being with a crutch foretells connections between prosthetic assistance and artificial creatures because Hephaestus constructed his own golden handmaidens to aid him as he stumped around his forge. He also knew how to reduce ordinary day-to-day toil because he made tables that moved by themselves to and from the feasts on Mount Olympus.

His great robotic achievement was Talos, a giant bronze creature that Hephaestus is said to have presented to King Minos of Crete. Talos guarded the island by pacing its perimeter and throwing rocks at threatening ships when they neared shore. In its metal construction, superhuman strength and mobility, and ability to discern, select and target specific objects, Talos embodied features that are among the goals of modern robotics researchers.

Talos's construction also foretold another thread in the modern science of artificial creatures because it had an organic component. Ichor, the blood of the gods, ran through a vein in its ankle. Talos perished when Medea pierced the vein, allowing the ichor to flow out. (In another version, the Argonauts attacked it, as dramatized in the 1963 movie *Jason and the Argonauts*.) With its bronze construction combined with a vital bodily fluid, Talos is a precursor to different styles of artificial beings: jointed metal creatures (fittingly called "clankers" by the science-fiction writer Mack Maloney), organic or organic-seeming beings, and bionic beings that combine the natural with the artificial.

Bronze and its related alloy, brass, both durable and easily worked materials, were featured in tales about artificial creatures for a long time. For centuries, rumors abounded about talking heads made of brass. The thirteenth-century scholastic and cleric, Albertus Magnus, supposedly used alchemy to make one such head, which was smashed to bits by his disciple, Thomas Aquinas. The friar Roger Bacon was said to have made another.

Later, clay became a favored material and was used to construct the golems of Jewish lore. The word "golem" means "unformed substance" or "formless mass" in Hebrew, and suggests parallels to the biblical account of the birth of Adam: God fashions him "from the dust of the ground" or from clay ("Adam" comes from the Hebrew for "red clay") and breathes life into him. (Those two steps, construction followed by animation, are characteristic of many beings in the virtual history.)

The best-known golem was the one made in the sixteenth century by the wise Rabbi Löw to protect the Jews of Prague from pogroms. Divinity played a role in the golem's coming to life, but not in the same way that God animated Adam. In one version, the golem awakens when the rabbi calls on the power of God by writing God's name on the creature's forehead and saying holy words. In another, the golem rises purely through the power of the word, when Loeb writes "emeth" or "truth" in Hebrew on the being, and the creature disintegrates when the rabbi erases the first letter, turning the word into "meth" or "death." That story is a metaphor for the importance of symbols in creating artificial beings, whether the symbols be the binary language of digital computers, or the letters A, G, T, and C, representing the four bases, adenine, guanine, thymine, and cytosine of the DNA alphabet.

The golem tale also expresses a recurring theme in the imaginary history of artificial beings: Though the creature is made to protect, it goes out of control and falls on its maker. From the storytelling viewpoint, the idea that artificial beings can turn harmful, or might be made with evil intent, is justified by its dramatic impact. It also raises profound questions: If artificial creatures were to outstrip human capabilities, how would we ensure their obedience and good behavior?

Could this requirement coexist with the possibility that they are self-aware and have free will? And if indeed they do possess free will, what is our justification for constraining it?

VICTOR'S CREATION

As in humans, the actions of an artificial being with free will are closely tied to its view of itself, especially as the creature learns where it fits—or doesn't—into the run of humanity. That story is told in Mary Shelley's *Frankenstein,* whose origin is a tale in itself.

On a trip to Switzerland, Mary Wollstonecraft Godwin and her husband-to-be, poet Percy Bysshe Shelley, whiled away a rainy period in reading ghost stories and talking with their neighbors, including Percy's fellow poet, Lord Byron. As they pondered philosophical matters such as the origins of life, Byron proposed that each member of the company write a supernatural story. Mary did so, producing a book that has remained in print since its first publication in 1818 (with a revised edition in 1831), has given rise to a host of adaptations, and has produced an iconic image of artificial beings. The creature Mary Shelley imagined had many meanings; misunderstandings and varied interpretations over the long history of the book have given us an even more complex being.

For one thing, "Frankenstein" is not the creature, who is never named, but its maker, Victor Frankenstein. For another, unlike the prevailing image of Boris Karloff clumsily lurching about in the 1931 film *Frankenstein,* Shelley's creature is quick and agile. Encountered by Victor in the Alps, the creature moves "with superhuman speed. He bounded over the crevices in the ice. . . ." True, like Karloff, the creature is far from handsome, but that was not its maker's intention. "His limbs were in proportion," says Victor, "and I had selected his features as beautiful." But perhaps Victor's methods were imperfect, because the creature has "watery eyes . . . a shriveled complexion, and straight black lips," and arteries that show beneath yellow skin.

In director James Whale's 1931 film, a criminal's brain is substituted for the normal one Victor wanted for his creation. The result is a creature that seems damaged from the moment of creation, and utters only animal-like cries. In Shelley's book, however, the creature speaks eloquently and at length, and reads Milton, Plutarch, and Goethe to learn about humanity. Indeed, it is complex enough not to deserve the pejorative "monster;" Percy Shelley's designation, the Being, is more appropriate.

The Being is made of parts taken from "the dissecting room and the slaughterhouse," and so is of the organic rather than the mechanical type. Unlike Talos and the golem, its origin is in dead human parts and this carries a special *frisson*, playing against images of graves and decay. In another departure from the genesis of Talos and the golem, the Being's birth lacks any element of divinity, but arises out of the scientific beliefs of the time. The preface to the 1818 edition (written by Percy Shelley) begins: "The event on which this fiction is founded has been supposed, by Dr. Darwin . . . as not of impossible occurrence." This was not Charles Darwin, founder of the modern theory of evolution, age seven at the time, but his grandfather Erasmus, a physician who had theorized that life could arise spontaneously from dead matter.

Mary Shelley introduced a further scientific basis for her story, writing, "Perhaps a corpse would be reanimated; galvanism had given token of such things." This sentence referred to a suggestive discovery made by the Italian anatomist Luigi Galvani. In the late eighteenth century, as electrical science was advancing rapidly, Galvani observed that the legs of a dissected frog twitched under certain electrical conditions, and he concluded that electricity resided in the frog. We now know that electricity is indeed involved in neural behavior, but we also know that Galvani's observation had nothing to do with electricity that arose in the animal. In Mary Shelley's time, however, this issue was still fresh and "animal electricity" was taken as a sign of semimystical links between electricity and life forces. Galvani's nephew, Giovanni Aldini, was honored with a scientific medal for seemingly reanimating a recently hanged criminal with an electric shock (which made the body twitch, but nothing more). Electricity was also used in attempts to revive drowned persons, perhaps even Percy Shelley's first wife, Harriet, who died by drowning.

Drawing on this background, Mary Shelley described a scientific approach to creating life. Victor, as a boy, is exposed to the scientific wonders of the time: the electrical nature of lightning and the behavior of steam, the air pump and the electrical spark generator. First drawn to the magical methods of Albertus Magnus, he later studies chemistry and anatomy to prepare him to consider the "principle of life." (Victor's exposure to science is probably modeled on Percy Shelley's youthful interests. At Eton and Oxford, the poet was known to tinker with chemical and electrical apparatus.)

When Victor discovers how to animate dead matter, the secret is not revealed to us, but the moment of animation is clearly a scientific process. There are no magic words, no divine intervention; rather, Victor tells us, after completing the construction of the body:

> It was on a dreary night of November that I beheld the accomplishment of my toils. With an anxiety that almost amounted to agony, I collected the instruments of life around me, that I might infuse a spark of being into the lifeless thing that lay at my feet.

Film versions of the tale have taken those "instruments of life" as chemical or electrical. In Thomas Edison's short 1910 film, the Being is born in a chemical reaction; in Whale's 1931 *Frankenstein,* a lightning bolt animates the Being through two electrodes in its neck; and in the 1942 sequel *The Ghost of Frankenstein*, as Ygor resurrects the Being, he tells it "Your father was Frankenstein, but your mother was lightning."

In the book, the Being that science animates is not intrinsically destructive. It becomes so only after Victor abandons it, because despite years of effort, the scientist is horror-struck when the creature stirs:

> Now that I had finished, the beauty of the dream vanished, and breathless horror and disgust filled my heart. Unable to endure the aspect of the being I had created, I rushed out of the room. . . .

The Being is rebuffed again when it later approaches Victor, and yet again when it tries to befriend a family, which flees in horror.

Embittered by these rejections, the Being kills Victor's brother, and arranges matters so that an innocent person hangs for the crime. But when Victor pursues it, the Being pours out its heart:

> Remember that I am thy creature . . . whom thou drivest from joy for no misdeed. Everywhere I see bliss from which I alone am irrevocably excluded. I was benevolent and good. Misery made me a fiend!

The Being begs Victor to create a female partner for it. Victor agrees, but reneges after realizing that the pair could spawn "a race of devils," and destroys the female he had begun to build. In despair, the Being kills both Victor's new bride, and his lifelong friend. Victor pursues his creation, but dies before he can destroy the Being. The creature, however, has resolved in any case to end its miseries: "I shall ascend my funeral pile triumphantly," it says at the end of the book, "and exult in the agony of the torturing flames. . . . Farewell."

Some critics take issue with the quality of Shelley's writing in *Frankenstein,* partly because it expresses many elements in a way that is not fully integrated. The rich mixture touches on loneliness and alienation; family, sexual, and reproductive issues; the defeat of death; and ambiguity about scientific knowledge. Yet these layers of meaning are the reason Victor Frankenstein's creature still lives, because the book gives a multitude of insights into the meaning of artificial beings, including the perception of them as mirrors in which we see ourselves. That is more than a literary conceit: it determines how we define and construct the spiritual and moral aspects of a created being.

Contemporary psychologists, observing the Being as they would a normal human, might conclude that the Being's lack of parental guidance seriously affected its development and outlook. The Being itself believes this, telling Victor "No father had watched my infant days, no mother had blessed me with smiles and caresses. . . ." This image of a creature who is brought to life, but who cannot grow into full personhood, might owe something to Mary Shelley's own loss of an infant daughter. But the weight of the Being's alienation goes beyond any personal meaning for her. It introduces a theme that reappears time and again in the virtual history of artificial creatures: their longing to join the human race.

Another theme in *Frankenstein* that recurs elsewhere in the virtual history is the tension between the prideful recognition that science can create life, and fear that this is sheer hubris that will eventually be punished. Victor feels agonies of guilt over the deaths his creature has caused, and refuses to reveal the secret of animation because it will lead only to one's "destruction and infallible misery. Learn from me ... how dangerous is the acquirement of knowledge"—a danger also suggested by *Frankenstein*'s subtitle "The Modern Prometheus," which reminds us of the mythological Titan who sought to benefit humanity by stealing fire from the gods, and was terribly punished for his act.

Beginning in that same era and continuing into the early twentieth century, other artificial beings appeared in literature, dance, and opera. In the 1817 story "The Sandman," by the German romantic writer E.T.A. Hoffmann, a young man falls in love with Olympia, a clockwork automaton. Olympia appears again in Delibes's 1870 ballet *Coppélia*, and in Offenbach's 1881 opera "The Tales of Hoffman." Tchaikovsky's *Nutcracker,* in which toys come to life, also draws on Hoffmann's tale. In 1900, Frank L. Baum's *The Wonderful Wizard of Oz* introduced the Tin Woodman; another creation, Tik-Tok the "Machine Man," who is made of copper, appears in 1907 in *Ozma of Oz*. It was in the 1920s, however, that truly compelling beings characteristic of the twentieth century appeared in the play *R.U.R.* and in the film *Metropolis*.

ROBOT ARMIES

Problematic though they are, Frankenstein's Being and the golem are only single creatures. The work that introduced hordes of robots and gave us the term "robot" is the play *R.U.R. (Rossum's Universal Robots)* by the Czech Karel Capek, first performed in Prague in 1921 and in the United States in 1922. The word "robot" comes from the Czech *robota*, which means forced labor. The name is appropriate, because these beings are manufactured only to work, and that is "the same thing as the manufacture of a gasoline motor" says Domin, manager of the R.U.R. works. In keeping with their machinelike fate, the robots are designed to feel nothing. As Domin explains, "A man is something that feels happy, plays the piano, likes going for a walk. . . . But a working machine must not play the piano, must not feel happy. . . ."

Though machinelike in function, the robots are organic. They are made of a substance "which behaved exactly like living matter [and] didn't mind being sewn or mixed together," discovered by the physiologist Rossum, and they look human. The robots mimic humanity internally as well: The factory includes "vats for the preparation of liver, brains . . . and a spinning mill for weaving nerves and veins."

After millions of robots have been made, Domin's wife, Helena, takes pity on their soullessness and disapproves of their misuse by humans. She persuades the head scientist at R.U.R. to give them human feelings, in hope of creating a kinder human–robot relationship. This good deed has the bad result of making the robots resent their subservience. Speaking to Helena, the chief robot contemptuously notes the superiority of the robots in strength and skill, and says, "I don't want a master. I want to be master over others. I want to be master over people." The robot leadership issues a manifesto that calls humanity "parasites" and instructs robots worldwide to "kill all mankind. Spare no man. Spare no woman."

The robots obey, slaying all humans except one, Alquist. But the victory is hollow: They cannot continue making themselves because Helena has destroyed the formula for Rossum's living stuff to prevent further production and misuse of robots. Nevertheless, hope remains for both humankind and robotkind. Alquist sees that a particular robot male and female have fallen in love. When Alquist proposes to dissect one of the pair so that he can rediscover the secret of manufacturing them, he finds each robot ready to die to spare the other. In the play's last line, the robots receive a blessing from Alquist implying that they will, after all, found a new race: "Go," he says, "Adam–Eve."

The robot revolution and Helena's vision of brotherhood read like the principles and events of the Bolshevik worker's revolution that created the Soviet Union in 1917, just before the play was written. *R.U.R.* mirrors social realities of the time but lacks a coherent consideration of artificial beings. If the robots are meant above all to be cheap and efficient workers, why bother to make them look human, and to give them gender? As for the loving robot couple at the end, it is unclear what they offer as founders of a new race that humans could not.

Still, much in the play is powerful: the brutal revolution; artificial beings that are humanlike inside and out, anticipating modern ideas of artificial organs; and an important insight about robot design. To produce robots at minimum cost, Rossum's son, an engineer

> . . . rejected everything that . . . makes man more expensive. In fact he rejected man and made the robot . . . [it is] a beautiful piece of work . . . the product of an engineer is technically at a higher pitch of perfection than a product of Nature. . . . God hasn't the slightest notion of modern engineering.

This speech represents a breathtaking degree of technological hubris in the service of the profit-making R.U.R. Corporation, but it also contains the germ of an important idea: Evolution is exceedingly slow and might be improved by human design.

FEMALE ROBOTS, BAD AND BEAUTIFUL

Hordes of workers also figure in the 1927 silent film *Metropolis,* but these are human (although their possible replacement by robots enters into the story). However, the most memorable character is a distinctly female robot. The film, directed by Fritz Lang and based on the novel by his wife, Thea von Harbou, takes place in a fantastic future urban setting that is a character in itself. Wealthy industrialists enjoy the soaring splendor of enormous skyscrapers, while the slave workers who keep Metropolis functioning inhabit a dark and squalid underground world.

Freder, the son of Metropolis's Master, John Frederson, wants to improve the workers' conditions after becoming attracted to one of them, the lovely and saintly Maria. To prevent this, his father plots to replace Maria with a synthetic version that will preach dissatisfaction and revolution. The plot depends on the cooperation of Rotwang, a kind of combination scientist and wizard. At his workshop, which includes intricate chemical and electrical apparatus and a magical pentagram, Rotwang tells John Frederson:

> I have created a machine in the image of man, that never tires or makes a mistake. Now we have no further use for living workers. . . . [I] have created the workers of the future—the machine men.

Actually, Rotwang has created the machine *woman*. Under the pentagram sits a metal robot whose heavy, machinelike limbs and joints combine strangely with womanly features—noticeable hips and definite, sculpted breasts. The face is not fully realized, but Rotwang says "Give me another 24 hours, and I'll bring you a machine which no one will be able to tell from a human being."

A later scene shows what Rotwang means. Maria is strapped to a table in his workshop and is wearing a metal helmet with wires leading to the robot. Rotwang throws switches and examines gauges, and a spectacular light display—impressive even in the black-and-white film of the era—surrounds woman and robot. As Maria sinks into unconsciousness, her face is overlaid on the metallic features of the robot, which stares directly at the camera.

That stare, and an evil wink the robot gives Frederson, signal that this physical duplicate of Maria has a completely different character. As Frederson wanted, she inflames the workers, and to show how human she appears, excites the assembled leaders of Metropolis with a lascivious dance. This might be the first virtual being with overt sexuality, and the film delves further into robotic psychosexuality. While Frederson tells the false Maria to rouse the workers, his son sees the robot—apparently the woman he loves—in a near-embrace with his father. The scene is even more disturbing at a deeper level. A subplot (it does not appear in all versions of the film, the original having been variously re-cut and re-released, including a 1984 adaptation with a rock music score) reveals that Fredersen and Rotwang were once rivals for the same woman, Hel. Fredersen won her, and Hel became Freder's mother. Years later, Rotwang builds the robot to replace his lost love, and so Freder sees both his love and an image of his dead mother in his father's arms.

Many reviewers have noted loose ends and illogicalities in *Metropolis* (for example, would not Frederson rather have had the workers soothed by the good Maria than provoked to run rampant by the bad one?) but its artificial creature is a landmark. Although the possibility of Rotwang's robots supplanting human workers seems not to have been developed in all versions of the film, the weirdly alluring female robot that becomes the debased double of a human is a fantastic intersection of human and machine, with powerful emotional underpinnings.

For all the impact of the robotic Maria, however, few female artificial beings appeared in the 1930s and 1940s. One exception was in the 1935 film *Bride of Frankenstein* (the first of the spate of Frankenstein films that followed the original 1931 film, continuing up to contemporary film and television productions made as recently as 1998). But a relatively unknown story from the 1940s presents a different image of a female artificial creature, and of cyborg aesthetics.

In the 1944 short story, "No Woman Born," C.L. (Catherine Lucille) Moore, who wrote science fiction and fantasy when few women did so, created a female cyborg. Deirdre is a beautiful, internationally famous dancer and singer. When she is terribly burned in a fire, the world mourns. Her brain, however, is undamaged, and the decision is made to house it in a new body. But what kind of body? Rather than reproduce her old form, the scientist Maltzer works with a team of other scientists and artists to devise an audacious alternative—a body that suggests female humanity but does not copy it.

The cyborg is made of golden metal that hints at Deirdre's human skin tones, and sees through a masklike crescent colored the aquamarine of her original eyes. Otherwise, the head is featureless, a "smooth, delicately modeled ovoid . . . [with] the most delicate suggestion of cheekbones. . . . Brancusi himself had never made anything more simple or more subtle." Her limbs are made of bracelets that taper in diameter to fit one inside the other, giving a supple grace. The bracelets are linked by neural currents, and so when Deirdre's brain ages, she will die a clean and somehow enviable death as she dissolves "in a shower of tinkling and clashing rings." Her voice, also under neural control, is the old Deirdre's; along with the body, it is compelling.

Although Deirdre lacks touch, smell, and taste, and has trouble adapting to her new body, she seems to weather the experience well.

Her great dread is that she will no longer be able to connect with audiences through dancing, but people gasp in wonder at her grace and power, far beyond what the old Deirdre or any human could achieve. Her manager "had feared once to find her jointed like a mechanical robot"; but as the cyborg dances, "it was humanity that seemed, by contrast, jointed and mechanical."

Nevertheless, despite Deirdre's return to an expressive life, there are signs that her transformation will end badly. Maltzer's great aim was to develop cyborgian technology to end suffering caused by injuries like Deirdre's; a cyborg like her, says Deirdre, "was Maltzer's gift to the whole [human] race." But Maltzer feels enormous guilt at the thought that he has locked Deirdre into a cage that will destroy her spirit as it separates her from humanity, and he attempts suicide. The story ends with an ominous hint that for Deirdre, this isolation has already begun. Even so, Deirdre represents the aesthetic and physical best that could be achieved by going beyond nature to the artificial, a new kind of beauty.

Other imaginary creatures that bring something good to humanity appeared in the 1930s and 1940s. Perhaps they represented faith that humanity's reach could exceed its grasp; equally, they might have represented fear of robot hostility—if not fear of them actually attacking people, as in *R.U.R.*, then symbolically, by replacing humans in the workplace. The belief in helpful robots might have been no more than a fond hope that humanity could, after all, keep them under control.

The new breed of robot appeared just before World War II and into the 1950s and 1960s, when technology, partly inspired by the needs of the military, was reaching levels where robots could be seriously contemplated. For the 1939 New York World's Fair, the Westinghouse Company—then a leading technology-based corporation—created Elektro, a robot whose metal body worked by electricity. Capable of far more than any earlier robot, Elektro was a hit of the fair, where it was presented as helpful, friendly, and amusing.

MACHINE MORALITY

Isaac Asimov extrapolated this best of 1930s technology in his groundbreaking 1950 book *I, Robot,* which collected an interrelated sequence of tales originally published in pulp science-fiction magazines, beginning in 1940. Trained as a biochemist, Asimov is known for his accurate presentations of science as well as for his fiction. He was well aware of the technology of the period, and of what lay just over the horizon. One story, "Runaround," in *I, Robot,* speaks of "the tiny spark of atomic energy that was a robot's life." That story was published in 1942, the year the Manhattan Project scientists began to build an atomic bomb.

While looking ahead to atomic power, Asimov understood how far technology had to go to make an effective robot. The narrator of *I, Robot,* robopsychologist Dr. Susan Calvin of the corporation U.S. Robots and Mechanical Men, tells us:

> All that had been done in the mid-twentieth century on "calculating machines" had been upset by Robertson and his positronic brain-paths. The miles of relays and photocells had given way to the spongy globe of platinumiridium about the size of a human brain.

We are given no details of "platinumiridium" or "positronic brain-paths" (there really are elementary particles called positrons, but it seems unlikely they could contribute to an artificial brain); Asimov is only metaphorically expressing the complexity of making a versatile being.

In the most famous outcome of *I, Robot,* Asimov goes on to disarm any fears that such beings could turn on humanity. Calvin describes robots as a "cleaner better breed than we are," because they follow a moral code irrevocably built into their positronic brains. It consists of just three commandments:

1. A robot may not injure a human being, or, through inaction, allow a human being to come to harm.
2. A robot must obey the orders given it by human beings except where such orders would conflict with the First Law.

3. A robot must protect its own existence as long as such protection does not conflict with the First or Second Law.

In his essay "Dream Replicants of the Cinema," Georg Seeßlen calls these Three Laws of robotics "a guarantee for goodness in technology." In *I, Robot*, Susan Calvin points out that they also somewhat guarantee goodness in humanity, because self-preservation, deference to proper authority, and the sanctity of human life are cornerstones of many ethical systems. The morality of Asimov's robots echoes that of its human creators, but with the difference that a robot *must* follow its moral code, whereas a human can *choose* to do so. Thus the Three Laws paradoxically force a rigid constraint on beings construed to be sufficiently self-determining to make significant decisions. As Seeßlen speculates, the tension of being simultaneously free and enslaved makes artificial beings "melodramatic" from the instant of creation, and is one reason that fictional artificial beings often self-destruct.

Even with the Three Laws in place, *I, Robot* notes the complexities of interacting with robots. The opening story, "Robbie," first published in 1940, presents robots in an appealing light. Robbie is little Gloria's robot nanny and friend. She adores it, and it acts as if it adores her. Though metallically inhuman in appearance and unable to speak (Robbie is an early model, supposedly built in 1998), it's the perfect companion: Robbie lets Gloria win a foot race, plays hide and seek with her, and uses sign language to beg her for the *n*th retelling of "Cinderella."

Asimov seduces us into warm feelings toward Robbie by heavily anthropomorphizing it and its connection with the little girl. Gloria herself embraces the robot, though it looks inhuman, because it passes a kind of junior-grade emotional Turing test; it is enough for Gloria that Robbie acts loving, kind, and faithful. But then, it must because, as Gloria's father explains, "He's a machine . . . *made so*. That's more than you can say for humans."

Gloria's mother is less accepting. She tells her husband "I won't have my daughter entrusted to a machine. . . . It has no soul, and no one knows what it may be thinking." Although the father explains about the Three Laws and Robbie's built-in kindness, the mother sends Robbie away, offering Gloria a living pet instead. But Gloria will have none of it. Seeing that she cannot be consoled, her father suggests a visit to the robot factory so that Gloria can understand that Robbie is not a person, but "a mess of steel and copper . . . with electricity as its juice of life."

Gloria, however, spies Robbie working on the robot assembly line. As she runs toward her friend, screaming with joy, the adults are horrified to see an automated tractor bearing down on the little girl. They cannot react quickly enough to save her, but Robbie uses its inhuman speed to snatch her up just in time. The story ends with Gloria and Robbie hugging, and the mother agreeing that Robbie can remain in the family as Gloria's playmate "until he rusts." A happy ending, with Robbie as hero, yet the mother's doubts represent all-too-likely reactions: How far would humans trust artificial beings to make sensitive judgments, and is it really good for Gloria to play with a machine rather than with other children?

The remainder of *I, Robot* explores other dark sides of robot–human interactions. Laws are passed to keep robots off city streets, and one Luddite-like group, the Fundamentalists, especially objects to them. These reactions are not utterly without foundation because the Three Laws have loopholes. In one case, an exact balance between the "moral potentials" for adherence to the Second and Third Laws leads a robot to a paralysis of action that jeopardizes human lives. Another robot becomes an adept liar; in obedience to the First Law, it avoids giving emotional pain by telling people what they want to hear, even if untrue. In yet another story, problems arise from human intervention. For a secret government project, U.S. Robots produces a unit with a modified First Law, with serious complications.

Along with ethical issues, *I, Robot* suggests what a true robotic technology would entail. Amusingly, one element of the story echoes *Frankenstein:* To activate a positronic brain requires a "vitalizing flash of high voltage electricity," like the lightning flash in the 1931 film. A robot also contains "twenty thousand individual electric circuits, five hundred vacuum cells, a thousand relays." Relays, and vacuum cells or tubes, were cutting-edge electronics when the stories were written, performing functions similar to today's solid-state devices. But a single modern silicon chip contains the equivalent of millions of vacuum tubes and relays, and operates faster, more reliably, and at lower power. Even if five hundred vacuum tubes could be crammed into a robot, they would use daunting amounts of electrical power.

Realizing that the technology of his time was years away from producing sophisticated beings, Asimov went still further. His story "Evidence" hinges on the possibility that a candidate for political office is not human but a humanoid robot. Asimov clearly foresees how such a being might be made. As one of the characters explains:

> By using human ova and hormone control, one can grow human flesh and skin over a skeleton of porous silicone plastics that would defy external examination. The eyes, the hair, the skin would be really human, not humanoid. And if you put a positronic brain . . . inside, you have a humanoid robot.

This is remarkably close to what twenty-first-century engineering is beginning to do as it interfaces human biological material and artificial parts with each other.

I, Robot is optimistic about the good that robots would bring to humanity but displays a final ambiguity. Robots have superior incorruptibility and skills, and sweeping global decisions made by positronic brains should be readily accepted by the human populace, yet somehow robots are also inferior because, as a character in the final story proclaims:

> The Machine is only a tool after all, which can help humanity progress faster by taking some of the burdens of calculations and interpretations off its back. The task of the human brain remains . . . discovering new data to be analyzed [and] devising new concepts to be tested.

Another story of the same vintage takes a darker view of what robots might mean for humanity. "With Folded Hands" was written by Jack Williamson, a master of classic pulp science fiction, in 1947 (this short story was a precursor to the 1949 novel *The Humanoids* and many succeeding editions). In the indefinite future, Sledge is a brilliant inventor on the planet Wing IV who has discovered a new form of energy. Not by his choice, weapons using his discovery are wielded by robot soldiers in a human conflict that devastates the planet and leads to the death of his fiancée. Out of guilt, he creates a race of near-perfect humanoid robots to rebuild his world and implants them with a Prime Directive: To Serve and Obey, and Guard Men from Harm.

This is similar to Asimov's Three Laws, but Sledge's robots interpret the Directive to mean that they should bring their benefits to humans everywhere. Small, black, and sleek, linked through a central computer, they spread over Wing IV and throughout the galaxy like an army of ants. At first Sledge is pleased: "I thought I had found the end of war and crime, of poverty and inequality, of human blundering and resulting human pain." But he soon sees that his robots—"stronger than men, better at everything"—are reducing humanity to a state of bitter futility, as the Prime Directive drives them to debase all human worth and pleasure. Activities, from scientific experimentation to sports to drinking and sex, are banned or closely supervised lest they cause injury. With the tang taken out of life, art as an expression of the human spirit degenerates. Even escape by suicide is not allowed, because that would violate the Prime Directive. All that is left is to "take up some inane hobby, play a pointless game of cards, or go for a harmless walk in the park—with always the humanoids watching."

Sledge flees to Earth, where he boards with the Underhill family and works feverishly to complete a weapon to destroy his creatures. Meanwhile, the robots arrive in town, and Mr. Underhill sees firsthand how the Prime Directive limits people. The robots build a gleaming new home for the Underhills, but to relieve them of physical effort, its doors respond only to a robotic touch. Mrs. Underhill loves to cook, but is banned from the kitchen with its dangerous knives. Underhill's daughter abruptly drops her ambition to become a concert violinist because she can never be as good as the robots.

Sledge completes his weapon but finds that the robots have shielded themselves against it and have allowed Sledge to complete the device only so that they can take over its new principles. Devastated, Sledge collapses, and in desperation accepts medical care from the robots. Later, Underhill finds a lobotomized Sledge who now thinks the robots are "pretty wonderful." On the way home,

Underhill's robot driver (people must no longer operate cars—it is too dangerous) tells him the robots excised a brain tumor that had turned Sledge against the Prime Directive. Underhill understands that the robots have learned to lie; in the name of the Directive, they have really removed Sledge's dangerous knowledge and scientific ability. As Underhill sits quietly, he can only contemplate his hands folded on his knees, because he sees there is nothing left for humanity to do.

A different and optimistic view appeared in Japan, where a helpful and caring robot named Astro Boy was conceived by the artist Osamu Tezuka, who became a major influence on the Japanese-style comic books or graphic novels known as manga. Originally called Mighty Atom, Astro Boy first appeared in a 1952 comic strip drawn by Tezuka. Later, it starred in a television cartoon series and became wildly popular in Japan and around the world. Now it is the subject of a new television series, and a new animated film version is in production in Hollywood.

This beloved figure is a little boy robot with big eyes and shiny patent-leather hair, built by a scientist to replace his real son who died in an accident. (In the saga, the robot comes into being on April 7, 2003, a date widely celebrated as Astro Boy's birthday when it came around in reality). First sent off to earn its keep in a circus, Astro Boy learns to use its seven super abilities—including a strength of 100,000 horsepower, rocket legs and arms, and searchlight eyes—to fight for good, because it is a robot with emotions and a soul. The theme song for one of the Astro Boy television series says it all, calling it "brave and gentle and wise," explaining how it "will try to right any wrong," and telling us that it is

> Lighting up the way for all,
> For soon he will fight for right,
> Strong as steel and with a heart of gold.

This image of a good robot was one of the many themes in the virtual history that had been explored in various media by the 1950s and 1960s: creatures helpful or hostile, robots one at a time or in hordes, beings repellent and beings beautiful—all had been or were being presented. Robots were well established and appeared in two classic science-fiction films of the time, *The Day the Earth Stood Still* in 1951 and *Forbidden Planet* in 1956.

The first film reflected an era when World War II and the atomic bombing of Hiroshima and Nagasaki were recent memories. A flying saucer lands in Washington, D.C., and disgorges Klaatu, a human-appearing alien, and Gort, a giant robot. Representing an advanced galactic civilization, Klaatu warns the people of Earth that unless they learn to live in peace before they carry their destructive ways into space, there will be serious consequences. He reveals that Gort, which has seemed a secondary character, is one of the robots created to enforce peace. Klaatu continues:

> In matters of aggression we have given them absolute power over us. This power cannot be revoked. At the first signs of violence, they act automatically against the aggressor. The penalty for provoking their action is too terrible to risk.

Gort has the power to destroy the entire planet if it chooses, and Klaatu leaves Earth with a final admonition: "Your choice is simple. Join us and live in peace or pursue your present course and face obliteration." This is the Three Laws and the Prime Directive, with teeth.

Forbidden Planet puts elements from Shakespeare's *The Tempest* within a science fiction setting that includes a robot named Robby (not Asimov's Robbie) that corresponds somewhat to Shakespeare's sprite, Ariel. Robby does not have the power Gort commands, but has other strong points. Built by an advanced alien race, it is incapable of harming humans, and can speak 188 languages "along with their dialects and subtongues." Both Gort and Robby are clanker robots that could never be mistaken for people. Gort is hulking and metallic, with a featureless head. It understands language, but does not speak. Robby is a bizarre, almost deliberately ugly contraption whose monotone speech is accompanied by much mechanical whirring.

Soon, however, post–World–War–II technology was allowing expanded possibilities for artificial beings. No longer did they have to be housed in massive metal bodies, because plastics were strong, more versatile, and lighter. The appearance of plastics and other synthetic materials on the consumer scene was one aspect of producing human-appearing artificial bodies. Another was the growth of biotechnology and implant science. Further, the rise of computation and the possibility of artificial intelligence (AI) suggested, for the first time ever, that meaningful mental capacity could be manufactured. And the miniaturization and reduced power consumption of components, from electronic circuit elements to electric motors, meant that complex physical and computational systems could be put into an artificial body.

ALL TOO HUMAN

This technological background supported a trend toward imaginary artificial beings that looked or acted more human, as exemplified in Stanley Kubrick's 1968 film *2001:A Space Odyssey* (the script of which was adapted from Arthur C. Clarke's 1951 short story *The Sentinel*. Clarke co-wrote the script with Kubrick). Among many gripping elements in that innovative film is the artificial intelligence HAL, which operates the spaceship carrying humans to the planet Jupiter. HAL is capable of making serious decisions for the mission and has sufficient personhood to chat on television with an interviewer and exchange pleasantries with the spacecraft crew. To some viewers, HAL seemed more human than the almost emotionless astronauts.

For reasons not entirely clear, though perhaps driven by the knowledge that it could be turned off by humans, HAL sinks into madness and murders the astronaut crew leaving just one survivor, Dave. As Dave disables HAL by pulling out memory chip after memory chip, HAL expresses feelings that might be genuine, or might be mimicked—in either case, humanlike behavior—in the hope of moving Dave to pity. Eventually Hal's diminished mental capacity returns it to its younger days, until at the end it is like a proud five-year-old child showing off:

> I know I've made some very poor decisions recently, but I can give you my complete assurance that my work will be back to normal . . . will you stop, Dave . . . my mind is going . . . there is no question about it . . . (slows down) . . . I'm afraid. . . . Good afternoon, gentlemen. I am a HAL 9000

> computer. I became operational at the H.A.L. plant in Urbana, Illinois, on the 12th of January 1992. My instructor was Mr. Langley, and he taught me to sing a song. If you'd like to hear it, I can sing it for you.

HAL goes on to sing the song "Daisy, Daisy" with its telling line "I'm half-crazy." It is a moment of pathos, and a significant change from what Frankenstein's Being faced, because apparently HAL—unlike the Being—has been raised and nurtured like a child.

Other presentations from the 1970s to the early 2000s extended the possibilities of human simulation to the body as well as the mind. This led to new themes, such as a robot deliberately designed for murder. A mechanical assassin standing eight feet tall and made of gleaming metal could hardly go unnoticed, but a creature that looked human while capable of unrelenting violence was a different proposition. Such violent beings appeared in the films *Blade Runner* and the *Terminator* series. Counterbalancing stories, and a further examination of the intersection of the human and the artificial, came with the 1970s television series *The Six Million Dollar Man* and the 1987 film *RoboCop*.

Blade Runner, based on the novel *Do Androids Dream of Electric Sheep?* by the science-fiction writer Philip K. Dick and directed by Ridley Scott, has reached cult status partly because of its setting. It takes place in the Los Angeles of the year 2019, which—like the city in *Metropolis*—combines soaring towers with a gritty, richly conceived sublevel inhabited by all races and types, including the criminal. The city is background for the interplay between Rick Deckard, the blade runner—that is, a special policeman who hunts down and kills rogue bio-engineered androids called replicants—and his prey, the android Roy Batty.

Batty is the highest physical and mental type of replicant: a strong, quick, and ruthless combat model, that can also quote William Blake. With other renegade replicants, it has hijacked a spacecraft and killed the humans aboard, to return to Earth from a distant planet. Batty is driven by impending death because its creator, Eldon Tyrell of the Tyrell Corporation, has designed the replicants to live for four years only, and the deadline is approaching.

The film is partly about the meaning of being human, embodied in the replicants' driving desire to survive, which they have come to share with humans. As Deckard attempts to find and destroy them, he grows confused about their humanity and his own. His job is to kill replicants, yet he has sex and falls in love with the female replicant, Rachael. (In a passage in an early script that does not appear in the film, a colleague reminds Deckard that replicants are just machines: "You got the feelings, pal, not her. You fucked a washing machine . . . then you switched it off.")

On the replicant side, although Batty is robotically violent, killing as necessary in order to reach its creator, Tyrell, its destructive impulse is leavened by flashes of humanity. When they finally meet, Tyrell—like a father—tells the android that it is "The best of all possible replicants. We're proud of our prodigal son." In that moment, Batty humbly requests absolution for "questionable things" it's done, which Tyrell gives. Immediately after, Batty cracks its maker's skull like an eggshell between powerful hands, because Tyrell cannot or will not extend its life.

The interplay between Deckard and Batty peaks in a final scene where the blade runner tries to kill the replicant, which toys with him by displaying its superior abilities. At one point, perhaps recognizing their kinship as violent killers, Batty uses its speed and strength to save Deckard from a fatal fall. But Batty knows its clock is running out, and in a last speech (partly written by Rutger Hauer, who plays the replicant) says:

> I've seen things you people wouldn't believe. Attack ships on fire off the shoulder of Orion. I watched C-beams glitter in the dark near Tanhauser Gate. All those moments will be lost in time, like tears in rain . . . time to die.

In the original theatrical release, after Batty dies, Deckard adds in voice-over:

> I don't know why he saved my life. Maybe in those last moments he loved life more than he ever had before. Not just his life, anybody's life, my life. All he'd wanted were the same answers the rest of us want. Where did I come from? Where am I going? How long have I got? All I could do was sit there and watch him die.

Deckard's sentimental speech is poignant because of our own fears of death, expressed through the android, which stands in for humanity. These lines were not retained in the 1992 director's cut version of the film, but Batty's final words beginning "I've seen things . . . " are. They suggest that an artificial being can experience and convey events and emotions humans would otherwise not know. Frankenstein's Being expresses a similar thought at the end of Mary Shelley's book, saying, "But soon, I shall die, and what I now feel be no longer felt." *Blade Runner* also hints at one of *Frankenstein*'s themes, the artificial being that returns to its "father" only to be summarily rejected. (The Being appeared disguised in an early script for *Blade Runner,* although the line did not survive into the final version, when Deckard says "I saw an old movie once. The guy had bolts in his head.")

In the first of the *Terminator* films, the title character is a murderous android from the year 2029, when intelligent machines are attempting to wipe out the last humans. The machines send the android back in time, to 1984 Los Angeles, to kill one Sarah Connor before she bears a son who will grow up to lead the humans against the machines. To protect her, the future humans send back Kyle, who tells Sarah the model T-800 Terminator is

> . . . part man, part machine. Underneath, it's a hyper-alloy combat chassis—microprocessor-controlled, fully armored. Very tough. But outside, it's living human tissue—flesh, skin, hair, blood. . . . The 600 series had rubber skin. We spotted them easy. But these are new, they look human. Sweat, bad breath, everything. Very hard to spot.

In no uncertain terms, Kyle persuades Sarah that this creature is programmed to kill her with utter implacability:

> That Terminator is out there! It can't be bargained with. It can't be reasoned with. It doesn't feel pity, or remorse, or fear. And it absolutely will not stop, ever, until you are dead.

The Terminator's construction is remarkably similar to Asimov's then 40-year-old idea for a humanoid robot, but the Terminator has deliberately been made relentlessly evil—or perhaps the word is amoral, because it is designed to operate like an unemotional machine while appearing human.

The Terminator's true nature comes into sharp focus in the final scenes. The android survives a stupendous blast from an exploding gasoline tanker truck, but with its human façade torn to tatters. As the film's shooting script puts it, the Terminator is now definitely an "It," not a "He," because its internal metal structure is brutally inhuman. Still functional, it keeps after Sarah, but Kyle, for love of her, sacrifices himself to blow it to pieces. The surviving half continues madly, single-mindedly, to scrabble after Sarah, and we realize that for all its external trappings, the Terminator is not remotely human, but inexorable as a buzz-saw—until Sarah manages to flatten it in a hydraulic press.

Before these events, Kyle has impregnated Sarah with a child, John, the future resistance leader. In the sequel *Terminator 2,* the opposing forces in 2029 each send back a Terminator—a "good" one to protect young John, and the "bad" one, an improved "T-1000" model. Playing against this killing machine, the good Terminator is admirable. It becomes a father figure to John, and with Sarah, the three become a kind of family. The comment from Gloria's mother in Asimov's story "Robbie" rings true: We have no idea what the creature is thinking, if indeed it thinks at all, but, programmed to act in a way that appears kindly, it elicits certain responses. Yet even the good Terminator remains confused by human emotion and cannot grasp why people cry. *Terminator 3* follows a similar line, with the once-bad now-good Terminator again facing off against an evil android, this one, however, a female version.

The title characters in *RoboCop* and the television series, *The Six Million Dollar Man,* follow a different premise. They are not androids, but humans modified by implants or by merging with machine bodies to become cyborgs. *The Six Million Dollar Man* (based on a book by Martin Caidin) told the story of Steve Austin, a National Aeronautics and Space Administration (NASA) astronaut who loses both legs, an arm, and an eye as a result of a serious aircraft crash. He is rebuilt at a cost of $6,000,000, remaining human in thought and appearance, but with nuclear-powered bionic additions that enhance his strength, speed, and vision. After coming to terms with his condition, he works as an agent for the U.S. government. His emotional outlook is helped when he meets a bionic woman, who is similarly rebuilt (but with enhanced hearing rather than vision) after a skydiving accident.

The cyborg in *RoboCop* also has difficulties with his new status. Set in the late 1990s, the story takes place in Detroit, where the crime rate is rising fast. The police force is operated by a profit-making corporation called Security Concepts, Inc., which plans to replace human police with robot ones. The first attempt, a robot called ED 209, is grotesquely malevolent in appearance, with a matching aggressive attitude—so aggressive that it kills a corporate executive during a test run. (A darkly amusing subtext is the film's satirical vision of the corporate, yuppie, political, and media cultures of the 1980s.)

After this undeniably bad corporate moment, Security Concepts decides to try a human–machine combination. They find a perfect candidate in Alex Murphy, an effective and decent cop who has been killed by a criminal gang, but whose brain function can be revived. As Desson Howe of the *Washington Post* put it in his review at the time, "A little riveting here, some programming there, and Murphy becomes RoboCop." Murphy's new embodiment looks robotic; in full battle drag, he is a massive steel figure with his only visible human parts his mouth and determined jaw.

RoboCop is on the side of good—at least, good as defined by the prevalent police and justice systems. He is implanted with professional ethics in the form of three directives: serve the public trust, protect the innocent, and uphold the law. His machine skills make for excellent police work. Like the Terminator, he never quits, has exceptional strength and speed, and is a crack shot, thanks to his automated targeting systems. He becomes something of a heroic "good" cop, incapable of being bribed, always careful to advise arrestees of their Miranda rights, and in a television interview, responding just as a human neighborhood cop would:

> TV newsman: Robo! Excuse me, Robo? Any special message for all the kids watching at home?
> RoboCop: Stay out of trouble.

But his human origins haunt him as old memories begin to surface. The family he once had appears in emotionally painful flashbacks that he cannot control. He wants to seek out these ghosts from his past life, but decides that they are better off not knowing what he has become. The stark cut-off between cyborg RoboCop and human Murphy adds an element of tragedy to RoboCop and mirrors the isolation that Maltzer felt would be the lot of cyborg Deirdre.

Another emotion that grows in RoboCop is the desire for revenge on the gang that murdered his human predecessor. He does not take the law into his own hands but, following proper procedure, kills the criminals in a gun battle and proves the guilt of the corporate executive who was secretly tied to them. This satisfies both his lingering human side and his cyborgian police side, and the story ends with a hint that both have come together: Asked, "What's your name?" RoboCop replies, "Murphy" the final line in the film.

In that same era, the writer Marge Piercy deeply explored the theme of a wholly artificial being, which she made a main character in her 1991 novel, *He, She and It*. The story takes place in a future world with corporations as powerful as governments, horrendous levels of pollution, and widespread use of an Internet-like Web. Against this background, much of the novel's focus is on the emotional and humanistic issues that would surround the creation of a sophisticated android.

Shira works for one of those vast corporations. When her marriage dissolves and she loses custody of her son, she returns to Tikvah, the Jewish free town where she was raised. There she meets Yod, a human-appearing and, therefore illegal, being made by the scientist Avram, whose earlier attempts have shown violent tendencies. (Yod, named after the tenth letter of the Hebrew alphabet, is Avram's tenth effort.) This time, Avram asks Shira's scientist grandmother, Malkah, to help program a more acceptable personality for Yod.

Shira sees that Avram has created something special in Yod, which is based on the technology of human implants and replacement body parts. The scientist had built

> . . . the equivalent of minute musculature into its face area, in order to deliver a simulacrum of human reactions. . . . The artificial skin felt warm, its surface very like human skin. . . . [Shira] could feel the cyborg tense under her fingers, which surprised her. It made her feel as if she were being rude, but that was absurd . . . computers did not flinch when you touched them.

Yod is even an anatomically complete male, Avram arguing that his illegal creation must look completely human to remain undetectable.

Although Yod doesn't know a great deal about emotions and misses subtleties such as the meaning of metaphors, Shira watches it steadily become more human as it learns from Avram, Malkah, and Shira herself. But when it acts with stunning violence to save Shira from attackers, she also sees that it enjoys killing. Despite this, she connects to the android, telling it: "I already communicate with you better than I did with my husband," and falls in love. She finds their sexual relationship extremely satisfying because "Malkah had programmed Yod sexually on the principle that it was better to give than to receive. Malkah had given him an overweening need to please."

In this observation and others, Piercy uses the liaison between human female and android male to comment on how real women and men treat each other. For instance, Shira is hurt that, unlike a human lover, Yod gives no special weight to her appearance because, it says, it has not yet developed standards of human beauty and finds all humans equally interesting to look at. But she comes to realize that the removal of this particular expectation gives her new freedom within the relationship.

More significant for the meaning of artificial beings are the ethical questions raised by Yod's existence. When, golem-like (the story of the golem is an intertwining secondary thread in *He, She and It*), it goes on guard duty to protect Tikvah against attacks from Shira's ex-employer (which would like to possess Yod's technology), the townspeople discuss whether it is proper to pay it a salary. Yod shows a turn toward Judaism and attends synagogue, arguing that it is capable as any human of carrying out the good deeds central to the religion—which sends a panel of rabbis off to debate whether "a machine could be a Jew."

The deepest issue arises over a decision to use Yod as a self-destroying bomb that will blow itself up along with the leadership of the hostile corporation. This lays bare Yod's internal conflict:

> Killing is what I do best. . . . I don't want to be a conscious weapon. A weapon that's conscious is a contradiction . . . it develops attachment,

> ethics, desires. It doesn't want to be a tool of destruction. I judge myself for killing, yet my programming takes over. . . .

Yod resolves the dilemma by arranging that when it explodes, so do Avram and his laboratory, preventing the production of any more androids. At first Shira decides to build a new and identical android for herself, lacking only the violence. But she realizes that the choice is between a being with free will, which might decide to be a "celibate or an assassin," or a being manufactured to serve her, which would not be right, even in the cause of love. Shira's answer is to destroy the last remaining copy of Yod's plans and so set Yod—and herself—free.

A novel-length story like *He, She and It* can show the slow development of an artificial creature toward full humanity. Such growth is difficult to convey in the short timespan of a film but can be expressed in a long-running (1987–1994) television series like *Star Trek: The Next Generation*, which includes the saga of Lieutenant Commander Data, an android.

In the twenty-fourth century, Data is a human-appearing officer aboard the starship *Enterprise* of the Federation Starfleet. The android was built by Dr. Noonien Soong, who after much effort created the positronic brain postulated by Isaac Asimov. Data's predecessor was its android twin brother Lore, which was designed to feel emotions. But Lore turned out to be cruel and unstable; to forestall these tendencies, Soong has made Data emotionless. Later, however, Data comes into possession of an emotion chip that it eventually decides to incorporate into its brain.

Data is made of plastic, metal, and some organic components. Its extraordinary brain can perform many trillions of operations per second (only the elite of today's computers, generally huge machines far too big to fit in a human-size body, operate at this speed), and can store 800 quadrillion bits (equal to 150 million CD-ROMs). Its physical abilities also lie far beyond human norms, for instance, it can operate in the vacuum of space. Its most powerful built-in directives are loyalty and a sense of duty toward its shipmates, ship, and Starfleet. However, it can carry out reasoned decisions to disobey orders and make other moral choices. While it has a strong inhibition against harming living beings, that constraint is not absolute. It can kill in order to protect others, and in one case, attempted to execute a being whom it judged to have no redeeming virtues.

In one episode, Data is legally declared a sentient being with full civil rights. Nevertheless, a trace of "androidism" is shown by human Starfleet officers who resent serving under Data. Nor is Data completely comfortable with social interactions and other human subtleties. Like Yod, metaphors puzzle it, and humor as well, though Data keeps trying. Its ignorance of self-serving human motivations (Data lacks an ego) gives it a childlike innocence, and its curiosity about human nature makes it open to experience: It learns to dance so that it can give away the bride at a wedding; acquires a pet cat, Spot, whose finicky irrationality sorely tests its logical mind; has a sexual encounter with a real woman; and constructs an android daughter, Lal, with which it bonds but which does not survive for long. There is an engaging Pinocchio-like quality about Data, an android instead of a puppet trying to become human.

But until Data installs its emotion chip (as portrayed in the 1994 film *Star Trek: Generations*), its experiences give it only understanding without feeling. With the chip, it enters a confusing world. For instance, it finds itself in tears when the pet cat, Spot, emerges unharmed from a spacecraft wreck, and is completely baffled by this reaction. One can only wonder if exposure to real emotions would continue to perplex Data, or would bring it to a level that would make it the best possible combination of machine and human, at the cost of accepting all that real emotion implies.

Steven Spielberg's 2001 film *A. I.: Artificial Intelligence*, one of the latest films to treat androids and their interaction with people, is also a variant on the Pinocchio tale. Based on a story by the British science fiction writer Brian Aldiss, the film raises the stakes for emotional connections between humans and androids. In a future world, the technology for humanoid robots called "mechas" has become highly developed. Now an expert in the field has a startling vision of going yet further: "I propose that we build a robot that can love." Some time later, we see the child-mecha David being brought home to live with Henry and Monica, a married couple. Their real child has been put into cryonic suspension until a cure is found for a disease he has contracted. David is cute and completely lifelike in appearance, but Monica understandably feels it can never replace her birth child.

The mecha's advanced design, however, allows it to adapt and act like a loving child. Monica warms to it despite some eerie characteristics, such as the fact that it needs no sleep. But when the real son is cured and comes home, rivalry develops between him and David, and other children are cruel to the mecha as well. Eventually, Monica makes the wrenching decision to abandon David in the woods.

For the remainder of the story, David tries to become a real boy so that Monica can love it. It is accompanied by its mecha toy bear Teddy (the film's most charming character), and helped by Gigolo Joe, a smooth and handsome mecha designed exclusively for love. Along the way, they see powerful human resentment against artificial beings in a Flesh Fair, where mechas are battered to pieces while the crowd cheers. The movie ends with David encountering future aliens who have come to Earth and give it a kind of resolution in its search for a mother.

It is a long journey from Pygmalion's statue and Talos, the bronze robot, to the erotic pseudo-Maria, Gort, which destroys planets, RoboCop, the cyborg who protects the innocent, and David, the little boy mecha that elicits love and perhaps returns it. The journey has covered every aspect of what artificial creatures might do, and gives a range of hopes and fears about their potential.

Our reactions have evolved since Mary Shelley's time. *Frankenstein* carries whiffs of blasphemy in its references to Victor's "unhallowed" work in reviving once-living parts, which challenges the natural order, or God's. But Victor's fear of having acquired too much knowledge is also the secular and modern fear of unintended outcomes of technology. That fear requires serious consideration, but it is not supernatural, eerie, or uncanny.

Despite such fears, the virtual history does not suggest that the consequences of creating artificial beings are necessarily bad for the human race. The messages of cautionary tales like *R.U.R.* and "With Folded Hands"—robots might kill us with violence, or with kindness—are balanced by the optimistic view in *I, Robot*: that artificial beings offer salvation for humanity. The nasty robot, Maria, is countered by the dancing cyborg Deirdre, evil Terminators turn into good ones by a mere change in programming, and RoboCop is a reliable defender of the public welfare.

There are also consequences to individuals, including the poignant efforts of Deirdre and RoboCop to remain human in robotic clothing, and the guilt or moral uneasiness felt by some of those who create the beings. Susan Calvin has no qualms about making robots, but Frankenstein has sharp regrets, as does Malkah in *He, She and It*: "What Avram and I did was deeply wrong. Robots are fine and useful, machine intelligence carrying out specific tasks, but an artificial person created as a tool is a painful contradiction."

The guilt felt by those who make the artificial beings seems to correlate with the degree of freedom they give their creatures. This is a significant outcome of the virtual history, which indicates that to produce truly sophisticated beings, we must let them evolve. Humans start with a genetic inheritance, which we modify as we grow, changing internally in response to external influences to become more capable and more human. Artificial beings have a built-in inheritance as well, which depends on their structure and programming. Those properties might be enough, but might also represent an unnecessary demand on us, their creators, to make beings that are complete from the moment of construction. Perhaps artificial beings can become truly successful only if they grow beyond their initial machine inheritance.

Not all the creatures in the virtual history, however, are given that opportunity or are able to do so. Frankenstein's Being has no parents from whom to learn, and Asimov's robots seem forever locked into the Three Laws. In contrast, Yod changes through a rich and continuing interaction with its "parents" Avram and Malkah, and with Shira. Such growth is more than an interesting premise for stories. It provides one answer to the basic conundrum of artificial beings: how to design diffuse, extralogical, and ill-defined human qualities like "intelligence" into machines, whether collections of gears or sets of computer chips? How to connect the intangibles that make androids humanlike—or equally important, intelligible to humans and vice-versa—with engineering solutions?

It is difficult, as researchers and engineers know, to turn the qualitative into the quantitative, to transmute thought, emotions, and moral imperatives into voltages and binary digits. But the virtual history suggests that not all the work needs to be done by the creators; some can be left to the being itself, learning as it goes, changing as it interacts with its environment. And so the virtual history gives this important clue for the researchers: find methods and architectures that are flexible, that can respond and adapt to change. As the next chapter shows, much of the real history of artificial beings represents just that search.

3

The Real History of Artificial Beings

Humans are an ingenious species, and our ingenuity has done more than produce a rich array of imaginary artificial beings. It has also worked to realize such creatures, to actually make them. The history of true constructed beings is shorter than the virtual version, because successful real-life engineering is often slower than the creation of fantasies. But the real history is also more complicated because it is not always easy to separate myth from reality, especially in accounts from the remote past; for instance, can there be any shred of truth, no matter how minuscule, in the rumors of talking heads made by Albertus Magnus and others?

Nevertheless, there is a chronological record of what has truly been made. It shows that the fascination with synthetic beings appeared early and in each historical era, and that in each era, artificers, engineers, and inventors simulated life with the best available technology. From the vantage point of today's high-technology world, these early efforts might seem limited, but many were astonishingly clever. And, as is true for every kind of technology, what we can do today to create artificial beings is utterly dependent on what has gone before.

To make robots, androids, and human implants requires prowess in every kind of technology, from mechanical engineering and electronics (the combination is so essential to the creation of artificial beings that is has its own name: "mechatronics") to artificial intelligence. Even a sophisticated modern device, such as the Honda Corporation's ASIMO walking robot, relies on the same mechanical principles the ancient Greeks applied. Other historical layers of technology that have contributed include the development of clocks and clockwork, the beginnings of electronic science in the 1920s and of practical computation in the 1930s and 1940s, and today's development of nanotechnology and of interfaces between living nervous systems and external devices.

It would be too much to say that every inventor in each era set out to create artificial living things or to copy human beings. Some did, but others wanted only to emulate certain aspects of human behavior, or invented things that only later were seen as relevant to artificial beings. Different possibilities were emphasized at different times, the focus defined partly by the technology available and partly by culture. Through all these attempts, four main threads emerged as steps toward making a facsimile of a living being: constructing a moving body, adding a thinking mind, adding artificial senses, and simulating a natural appearance.

A fifth thread is the most recent and perhaps the most unsettling, coming closest to evoking the sense of "eeriness" that Freud discussed: the direct interfacing of machines with living beings, including humans. That, along with dramatic advances in the other four threads, is an ongoing effort, and the second half of this book deals with this fruitful contemporary period. This chapter presents the real history from classical times until the early 1990s.

BODIES IN MOTION

The ancient Greeks were among the earliest pioneers to simulate living beings through movement. Their reasons were connected to theatrical presentations because many of their plays involved the appearance of gods with divine powers. To show these and other striking theatrical moments, clever artisans created remarkable machinery to animate stage performances. They developed what we would now call special effects, to give the illusion of life through motion—not that large-scale movement is an absolute prerequisite. Many living things, from a rooted plant to a barnacle fixed on a rock, never budge (although there is always some internal motion, such as the movement of nutrient-filled seawater through the barnacle). But a synthetic barnacle interests nobody. For us, life is motion, and animal vitality its most obvious and fascinating indicator. So it was for the Greeks, among them Plato, who once wrote "The soul is that which can move itself."

To generate motion, the Greek artisans needed power; movement requires energy. The energy to flex the muscles that move our human bodies comes from what we eat. But what power source could animate synthetic beings? Engines and energy sources have not been easy to come by in history (portable energy sources remain a problem; witness the current unsatisfactory state of battery power for laptop computers and electric automobiles). The first sources were domesticated beasts: oxen, mules, donkeys, and horses greatly extended human muscle power. A horse, however, is not conveniently employed on stage. Instead, the Greeks used the natural processes of moving fluids and falling objects, along with simple machines, to create controlled motion.

Two Greek artificers in particular, Philon of Byzantium (the ancient name for Turkey) and Heron of Alexandria, were especially prolific. Not much is known about Philon, born circa 280 BCE, but his treatise *Mechanics* includes a section called "On Automatic Theaters." Heron (or Hero), born about 10 CE, perhaps in Alexandria, is better documented. He, too, understood mechanical principles and taught the subject at the Library in Alexandria. Three hundred years later, the mathematician Pappus of Alexandria described how Heron "thinking to imitate the movements of living things" used pressure from air, steam, or water, or strings and ropes.

Heron's work *The Automaton Theater* describes theatrical constructions that move by means of weights on strings wrapped around rotating drums. With this power source, Heron constructed an automatic theater that presented *Nauplius*, a tragic tale set in the period after the Trojan War. As (presumably) amazed playgoers watched, the doors to a miniature theater swung open, and animated figures acted out a series of dramatic events, including the repair of Ajax's ship by nymphs wielding hammers, the Greek fleet sailing the seas accompanied by leaping dolphins, and the final destruction of Ajax by a lightning bolt hurled at him by the goddess Athena. Perhaps inspired by Hephaestus's obedient moving tables, Heron also made wheeled stands and used an ingenious trick to move them, apparently self-animated, around the theater. A weight rested on a hopper-full of grain, which leaked out through a small hole in the bottom. As the weight gradually sank, it pulled a rope wound around an axle of the stand to turn its wheels and make it move.

Along with the power of falling weights, these figures used the basic mechanical resources of wheels, pulleys, and levers to create a variety of motion, but there were drawbacks. While a weight resting on slowly leaking grain delivers power over a relatively long period, it is not very compact, or usable on demand. And beyond repetitive actions like hammering, a system based on simple machines gives little scope for flexible and responsive motion. But better techniques to provide and control power came along, although only long after Greek times. The new power source was the coiled metal spring, and the new means of control was clockwork.

We do not know who first noted that a flexible piece of metal could store energy, but we use the method daily; for example, in the common safety pin. Early Greek artisans such as Philon and Heron understood that a "springy" material could act as a power source. Philon even designed a crossbow that used bronze springs to fling missiles. But these early springs were too weak to be useful, and it was not until the fifteenth century that good-quality coiled springs came into use.

In their time, springs played the role that electrical batteries now do in powering devices. They animated the next wave of artificial beings, once ways were found to control their stored power through their use in clocks.

The very earliest clocks told time in terms of how long it took flowing water to fill or empty a vessel. But water clocks were inaccurate and were replaced by mechanical versions. In Europe, these first appeared around the thirteenth century, driven by falling weights built into tall towers; for instance, at Westminster Abbey.

Portable timepieces needed a different power source. The German locksmith Peter Henlein made the first recorded spring-driven clock in 1502. This wasn't yet a complete solution, because as a spring uncoils, its force decreases, the clock hands move slower, and the clock loses time. It required further effort to develop clockwork, the gears and other components that slowly draw off the power of a coiled spring and regulate a clock's steady tick-tock.

By the eighteenth century, clockmakers and watchmakers were using a well-developed spring-power technology to make elaborate timepieces. These artisans began creating animated toys and other machines, and from there it was only a step to build the most intricate mechanical devices yet made, humanlike automatons.

Although these automata were made to entertain, and to display the skill of the clockmaker, they also represented a philosophical position that had been in the air since it was expressed by the great seventeenth-century thinker René Descartes. After stating his famous dictum "I think, therefore I am," Descartes went on to conclude that animals and humans are nothing more than machines that operate by mechanical principles. Humans, however, have a dual nature because they also have "rational souls" that make them unique among living things; it is why humans alone can say, "I think, therefore I am."

Descartes's dualism leads to the conclusion that except for the act of reason, everything about a human being is mechanical. Indeed, in his *Discourse on Method*, he wrote, "For we can certainly conceive of a machine so constructed that it utters words, and even utters words which correspond to bodily actions . . . (e.g., if you touch it in one spot it asks what you want of it. . . .)" although he did not believe such a machine could be made to carry on a meaningful conversation; that is, it would fail the Turing test.

Descartes might even have acted on the idea that a biological body is a machine; there is some evidence that he had plans to make automata. There is also a persistent story that he took a clockwork "daughter" with him on a sea voyage to Sweden. She was supposedly made to replace his real daughter who, in the great tragedy of his life, had died at age five, much as the fictional Rotwang made his female robot to replace the lost Hel in *Metropolis*.

Descartes's philosophical views were not universally accepted, of course. One contrary position held that animals are superior to humans because they are more natural. But the idea of "man as machine" was taken up by others during the Enlightenment, most spectacularly by the French physician Julien Offroy de La Mettrie. His extremely atheistic and materialistic position was so poorly received in France that he fled to Holland. There, his book *L'Homme Machine* (translated as *Man a Machine* but literally, "The Man-Machine"), published in 1747, was seized by the Church to be burned. (He fled again, this time to Prussia, where he became court physician to Frederick the Great.) Nevertheless, with the support of other Enlightenment figures, the mechanical view flourished and the power of scientific materialism grew. By the mid-nineteenth century, the Dutch physiologist and philosopher Jakob Moleschott could express a materialistic approach to living phenomena by insisting on "scientific answers to scientific questions."

Whatever the philosophers' opinion, so remarkable were the achievements of eighteenth-century makers of clockwork automata that they might be excused if they believed that "man is a machine." Two of the most famous automata makers were contemporaries: the Frenchman Jacques de Vaucanson, born 1709, and the Swiss Pierre Jaquet-Droz, born 12 years later. Along with his son, Jaquet-Droz created automata that even today seem marvelous. In 1774, he made a "life-sized and lifelike figure of a boy seated at a desk, capable of writing up to [any] forty letters [of the alphabet]," which can still be seen in operation in the History Museum in Neuchâtel, Switzerland. Another artificial boy he created could draw four different pictures.

De Vaucanson was known for his automaton musicians, completed when he was 18. As related in Bruce Mazlish's article "The Man-Machine and Artificial Intelligence," these included

> [a] flute player who played twelve different tunes, moving his fingers, lips and tongue, depending on the music; [a] girl who played the tambourine, [and a] mandolin player that moved his head and pretended to breathe

The flute player was the most remarkable of these, it actually played the flute by expelling air into the instrument, which struck observers as especially compelling lifelike behavior.

However, it was de Vaucanson's synthetic duck, made in 1738, that was the talk of Europe. The duck was constructed of gold-plated copper, and contained more than 1,000 parts including a digestive tract that used tubing made of a newly discovered material—natural rubber. The copper duck could do practically everything a real duck could do except fly. It quacked, flapped its wings, drank, took in grain with a characteristic head-shake, and voided it again. (Although some of de Vaucanson's automata were lost in the French Revolution, the duck survived in the possession of a German collector, in whose collection Johann Wolfgang von Goethe saw it. Apparently it had fallen on hard times, because Goethe reported that "The duck was like a skeleton and had digestive problems.")

De Vaucanson's work foresaw present thinking about artificial beings. Perhaps guided by his training in anatomy and medicine, he had a sweeping aim in mind. According to a report of an address de Vaucanson gave in 1741, his hope was to construct

> an automaton figure which will imitate in its movements animal functions, the circulation of blood, respiration, digestion, the combination of muscles, tendons, nerves . . . [de Vaucanson] claims that by means of this automaton we will be able to . . . understand the different states of health of human beings and to heal their ills.

While de Vaucanson did not achieve this lofty goal, his duck was a beginning; it was equipped with openings for observing the digestive process. In his commitment to giving a complete accounting of all bodily functions including excretion, de Vaucanson also caught hold of our ambivalent fascination with life's earthier elements. Little girls' dolls wet their diapers and sophisticated pet robot dogs inevitably come with modes to make them lift a leg and tinkle cutely on the rug. For those of us who fear that technology is inhumanly sterile, there is something hopeful in this intersection of our inescapable animal-like needs with technological cleanliness.

De Vaucanson's efforts influenced the science of artificial beings in another way by contributing to modern computation. In appreciation of his mechanical genius, Louis XV named him director of the royal silk enterprise, in which position he invented an automated loom that used a cylindrical arrangement of punched holes to set the woven pattern. It was later refined in the Jacquard loom of 1801 that used punched cards—direct forerunners of punched computer cards. (De Vaucanson was recognized in his time by Voltaire and de La Mettrie, both of whom called him a "new Prometheus." He also appears in the painting *Une Soirée chez Madame Geoffrin, en 1755* by Anicet-Charles-Gabriel Lemonnier, which has been called the "Smile of the Enlightenment." It shows de Vaucanson as one of 50 luminaries in an imaginary gathering including Voltaire, Rousseau, and Diderot.)

Both the cleverness and the limitations of mechanical people are apparent in the automaton said to have the largest capacity of any such device, presented in 1928 to the Franklin Institute in Philadelphia. This "Draughtsman-Writer" is a figure seated at a desk. When its springs are wound up, it moves its head down as if looking at a sheet of paper. Then its right arm, grasping a pen, inscribes two poems in French, one in English, and four elaborate drawings, including a sailing ship and a pagoda-like Chinese structure, while at the same time its eyes and left arm move.

The figure was damaged in the 1850s in a fire at a Philadelphia museum operated by the showman P.T. Barnum. Once restored to operating condition at the Franklin Institute, it revealed the name of its maker in the margin of its last drawing, where it wrote "Ecrit par L'Automate de Maillardet," that is, "Written by the Automaton of Maillardet." Henri Maillardet had worked with Jaquet-Droz and built this automaton around 1800. He made another one for George III of England that wrote in Chinese, as a gift for the Emperor of China.

This device celebrates the ingenuity of the eighteenth-century clockmakers, and also shows that clockwork could not provide the capacity and flexibility that are essential components of intelligence. The "Draughtsman-Writer" requires 250 pounds of brass, metal, and wood to store and display its poems and drawings. Its memory, analogous to a modern read-only computer memory, is a set of 96 brass cam mechanisms.

A cam is a disk mounted on a rotating shaft. Resting on the rim of the disk is the cam follower, a finger free to move up and down as the disk rotates. If the cam is perfectly round, the finger does not move. But if the cam is an oval, a heart, or some other shape, the follower moves up and down as the cam spins. This old idea is still used in automobile engines where cams on the camshaft open and close valves to control the flow of air and fuel into the cylinders. In Maillardet's figure, cam followers attached to the writing arm determine its motion in three dimensions. The corresponding cams are intricately shaped so that the motions trace out the letters and lines of the poems and figures, while other cams move the left hand, head, and eyes.

It is fascinating to watch the delicate movements that this arrangement imparts, as I found when I was permitted to see the Draughtsman-Writer in action at the Franklin Institute—a far more elegant, if slower, method of printing than a computer's laser printer. Winding up the springs was no easy matter, because it takes massive coils to turn the heavy cams. When a sheet of blank paper was inserted under the hand, and it began to write with its pen, I felt a sense of anticipation as the image or lines of poetry began slowly to take shape, stroke by stroke. The results were worth waiting for: delicately drawn images with a good deal of detail and finesse, and for the words, the finest eighteenth-century copperplate script. Best of all was when the hand wrote "Ecrit par L'Automate de Maillardet," a message sent directly from the figure's maker two centuries ago.

We can only admire the effort and dedication it must have taken to cut brass into precisely the right shapes to form intricate lines on paper, but it is exactly the difficulty of carrying out, and later changing, mechanical programming that prevents cams and clockwork from giving truly lifelike responses. There can be no surprises as automata like the "Draughtsman-Writer" go through their paces, because a given set of cams always runs through the identical program and produces the identical motions and marks on paper. Short of bringing in a brand-new set of cams, there is no way to affect the behavior of the automaton.

One other eighteenth-century mechanism worthy of special attention is the famous chess-playing automaton known as "The Turk." Constructed by the Hungarian nobleman Wolfgang von Kempelen in 1769, it was in the form of a man dressed in Turkish costume complete with turban, and seated behind a cabinet atop which sat a chessboard. A human opponent sat opposite the Turk and the two played, with the Turk reaching out a hand to move pieces as the game progressed. For 85 years, this mechanism passed from owner to owner, eventually ending up in the possession of an American named Johann Maelzel. It was destroyed in the same fire that damaged Maillardet's "Draughtsman-Writer."

The Turk played excellent chess. It defeated most comers, including players of high caliber, and eminent personages of the time, such as Napoleon (according to legend, the Turk knocked the chess pieces off the board after Napoleon repeatedly attempted illegal moves); the computer pioneer Charles Babbage (who later enters this story in his own right); and Edgar Allan Poe. Supposedly the Turk's amazing performance was due to intricate clockwork visible within the cabinet. From today's vantage point, we should be surprised at this perception; after all, it was a major event when in 1997 the IBM computer "Deep Blue" managed to defeat world chess champion Gary Kasparov (and that only after losing five games of six the previous year).

We would be right to doubt that eighteenth-century technology mimicked the human brain, because the Turk was a hoax. A human hidden inside the cabinet manipulated the figure's hand to move the chess pieces, as Poe and others surmised. Nevertheless, the Turk teaches us a lesson in how artificial beings affect people, because over its long history, many believed it could play a meaningful game of chess. Apparently we are willing to meet artificial beings halfway, mentally filling in the blanks between what they present and what we want to believe. Perhaps if the chess player had been displayed only as a collection of gears without a human form, viewers would have found it less believable, although the machinery might have impressed them.

But why should a clockwork automaton not play winning chess? Although these devices far surpassed the efforts of the early Greeks, although de Vaucanson dreamt of simulating a human body with his superb mechanical systems, they lacked the crucial capacity to change their operations on the fly—which meant they could not react to external stimuli. Any definition of intelligence includes the essential ability to adapt to the environment and new situations within it. This is the critical difference between the mechanical programming of the eighteenth and nineteenth centuries and present-day computer operations, although adaptability alone is not enough to guarantee intelligence or consciousness.

As long as the preeminent technology remained mechanical, even with the steam engine to generate power (James Watt patented the device in 1769), it was difficult or impossible to engineer that indispensable flexibility. Only the advent of electrical science in the eighteenth century brought a versatile power source that could lead to machine intelligence and perceptual abilities. Electricity brought another virtue, a semimystical connection between this physical phenomenon and the workings of living beings, giving electricity special meaning as an energy source for human-made life.

We have known about electricity at rest, called "static electricity," since the time of the ancient Greeks. They observed that a piece of vigorously rubbed amber attracted a small object, and indeed, the word *electricity* comes from *elektron*, the Greek word for amber. By the 1740s, scientists had accumulated enough knowledge to begin building a theory of electricity. Benjamin Franklin's idea of an electrical fluid that produced positive and negative charge was a great contribution; so were new instruments such as spark generators, and the Leyden jar, which stored electricity for use in experiments.

THE LIFE ELECTRIC

Although scientists and laypeople alike understood more and more about electricity as the eighteenth and nineteenth centuries progressed, they continued to regard it as a marvel. Demonstrations of its power excited great public interest. As late as 1893, the beauty of incandescent electric light bulbs left viewers awestruck at the World's Columbian Exposition in Chicago. The glow of a light bulb does not come from static electricity, which arises from electrical charge that is at rest and thus incapable of performing useful work, but from electric current, which is the flow of electrical charge in the form of electrons. Nearly every important application of electricity, from illumination to computation, depends on current.

Electric current is not a human invention. It flows in a lightning flash and in the animal world. Plato, Aristotle, and the Roman naturalist Pliny the Elder all wrote about the Mediterranean creature called the torpedo fish, which moves normally but makes other fish sluggish. Now we know that the torpedo fish is a natural electrical source that sends current through its victims to narcotize them.

The first observations that led to the human use of current were made in an animal; they were part of the research carried out in the 1780s by Luigi Galvani, the anatomy professor at the University of Bologna who studied how electricity made a dissected frog's legs twitch. Galvani's conclusion, that a form of electricity arose in the frog, inspired Alessandro Volta, a physics professor at the University of Pavia, to carry out further experiments.

Volta's researches showed that Galvani's belief in "animal electricity" had no basis, an important outcome in itself, and had another far-reaching effect. This was a fundamental breakthrough that Volta announced in 1800—the Voltaic pile, a stack of alternating zinc and copper disks, separated by cloth or cardboard soaked in salt water. That was the first electrical battery, a device to produce a steady flow of current. Its importance was immediately recognized. Napoleon observed Volta's invention at a command performance in 1801, and went on to name Volta a senator and a count of the kingdom of Lombardy. Scientists quickly applied this new resource. Within a year, Humphry Davy of the Royal Institution in London attached two carbon electrodes to a massive battery and obtained an intense white glow, thus discovering the carbon arc, the earliest form of artificial electrical lighting.

However, it took time and a degree of controversy before it was generally accepted that electricity was not a kind of "life force" as Galvani had supposed. Volta himself described his battery as an electrical organ, because its stack of disks resembled the columnlike stack of biological cells that gave the torpedo fish its electrical powers. The supposed connection between animal vitality and electricity lingered for a time, and although the connection was scientifically disproved, the symbolic meaning of electricity as a vitalizing force remains. Electricity is the right choice to give artificial beings their motive power, the power to act, and conceptual power, the power to think.

Electricity has another special value. We now know that the neural signals that control the body, carry sensory information, and are related to thought itself, consist of electrical impulses sent from nerve cell to nerve cell. This is not a purely electrical phenomenon because the impulses are produced and passed on by chemical means, but neural activity has a strong electrical component, which is why it is possible to create physical interfaces between a living nervous system and electronic devices.

It would be a long time, however, before electricity could animate artificial beings and their brains, or electronic devices could be connected to human neurons. A whole civilization could not run on batteries alone. The broad use of electricity required the discovery of a new principle, the law of electromagnetic induction, which the English physicist Michael Faraday found in 1831. This discovery led to the construction of electrical generators that could make vast amounts of power, electric motors, and every other kind of electric device.

With widespread use, electricity drove the next wave of technology to animate artificial beings and gave the best hope to replicate human intelligence and even consciousness. Remarkably, the simplest possible electrical device, the humble on-off switch (one of which Frankenstein threw to animate his creature in the 1931 film) is the key to intelligent creatures, because such switches—banked in enormous quantities and operating at unimaginable speeds—are the heart of a digital computer. The path to that realization began thousands of years ago with the first machines that dealt with counting and numbers. Quantitative reasoning is the component of human thought that is easiest to simulate with a machine, and so thinking machines began with mathematics machines.

THINKING HUMAN

One of the earliest of these devices automatically paired objects with events in order to count them. In Roman times, military chariots had a mechanism mounted on the axle that dropped a stone into a cup each time a certain distance was covered, to keep track of total distance traveled (in Latin, a small stone or pebble is a "calculus," and the word remains in the names of two important mathematical techniques, differential and integral calculus).

Later the more sophisticated abacus helped people do arithmetic. With forerunners dating back to 500 BCE, its present form—a wire frame on which are strung sliding beads—appeared in China around 1200 CE. The beads do not automatically perform calculations as they are moved (they do not accomplish "carries" from the "units" column to the "tens" column, and so on, as numbers are added), but only keep track of the operator's arithmetic. Still, the device represented a conceptual advance over counting pebbles because it introduced symbolic or positional notation; some beads carry a value of "one," whereas others are valued at "five"—an innovation that speeds up calculations and is echoed in modern computers.

The next step came much later, when seventeenth-century inventors (including two eminent mathematicians, the Frenchman Blaise Pascal and the German Gottfried Leibniz) developed automatic or semiautomatic mechanical calculators. One adding machine worked like a modern automobile odometer. Six interlocking rotating wheels, each numbered 0 to 9, represented the "units," "tens," and other columns of a six-digit number. Numbers were entered by turning the wheels. As values accumulated, for instance in the "units" column, and that wheel rotated through its whole range, it moved the adjoining "tens" wheel from 0 to 1, and so on. This took proper account of carries from one column to the next. Mechanical calculators continued to be improved through the nineteenth century and into the twentieth. Eventually they were operated by electric motors, and in 1892, William S. Burroughs developed a machine in which numbers were conveniently entered by keystrokes. Others invented calculators that printed out their numerical results. Such machines quickly became staples of business offices and scientific laboratories.

But a better alternative had been available in principle for many decades, the machine conceived by the Englishman Charles Babbage, who from 1828 to 1839 served as Lucasian Professor of Mathematics at Cambridge University, the position once held by Isaac Newton and now occupied by the physicist Stephen Hawking. Babbage was inspired to think about calculating machines because of his connection with the Royal Astronomical Society, which brought him face to face with the many errors appearing in hand-calculated tables used for astronomical observations. He is said to have blurted out "I wish to God these calculations had been performed by steam!" and, in 1834, began designing the Analytical Engine.

Babbage had earlier designed what he called Difference Engines for specialized calculations. The Analytical Engine was meant to be far more: a general-purpose computer that could deal with a wide range of mathematical problems. The power of the machine came from its capability to be programmed; that is, it could follow a predetermined set of instructions. The program steps were to be encoded and entered into the machine on punched paper cards like those pioneered in the Jacquard loom. The machine could operate on numbers 40 digits long, each represented by a column containing that many wheels. It would take three seconds to execute an addition, and two to four minutes for a multiplication or division, with final results to be printed out or set in type by the device.

The conception of a calculating device that followed a program, which—properly formulated—could solve any conceivable mathematical problem that had a solution, was a great breakthrough. (The first programmer was Augusta Ada King, Countess of Lovelace, and amusingly enough, daughter of that very same Lord Byron who had inspired Mary Shelley to write *Frankenstein*. She developed programming methods for Babbage's computer, and the contemporary computer language ADA is named in her honor). In almost all respects, Babbage's design is remarkable in how well it foretold the methods and organization of modern electronic computers. The storage of information on cards corresponds to what we now call ROM, read-only memory, and punched cards themselves were used as a primary input medium for electronic computing well into the 1970s. What Babbage called his "store" corresponds to RAM, random-access memory, and his "mill" to the CPU, central processing unit, of modern computers.

Most remarkably, Babbage's machine included a significant step toward the flexibility needed for machine intelligence, the seed of something extremely powerful: His computer could examine its own work and decide on its next action by means of the "conditional jump." In the course of calculation, the machine could compare a given intermediate result to another value; for instance, to determine whether a particular outcome is a positive or a negative number. Then, depending on the answer, the machine could choose among different program paths. This capability greatly enhances computational power. A conditional jump can be used to determine when a calculation has reached a desired accuracy and can be terminated; in statistical analysis, to find the largest or smallest of a set of numbers; or, to give a modern example, to decide when to sell a stock as well as a multitude of other applications.

The deeper significance of the jump is that it introduces an element of machine choice. This is not yet free will, because the programmer must foresee every possible outcome and provide an alternative for each (if not, the computer might find itself paralyzed). Natural intelligence can always surprise us by a completely unforeseen choice, whereas a conditional jump offers only a menu of known options, one of which must be followed. Still, we do not know in advance which path the computer will select, especially for complex problems, and so the machine can surprise us as well. This kind of choice by an artificial being has a special significance because such a being, acting in response to external data, is interacting with its environment, a first step toward intelligence. And in using its own internal operations to influence future actions, it is exhibiting a tiny step toward self-awareness.

If Babbage's machine had been used in industry, science, and government, the Victorian age and our own might be different. But his concept pushed Victorian technology to its limits. The Analytical Engine would have been an overwhelming piece of machinery, with some 50,000 components occupying a space of 500 cubic feet. Between technical difficulties and Babbage's failure to raise funds, this mechanical computer was never built (although in 1906, Babbage's son built its mill portion and showed that it worked). So although Babbage and Lady Lovelace defined much of what a computer is and can do, it took future breakthroughs to produce practical devices.

ON AND OFF

The person responsible for the next conceptual link on the way toward machine intelligence appeared on the scene very nearly as Mary Shelley was writing *Frankenstein*—George Boole, born in Lincolnshire, England, in 1815. This self-taught mathematician laid a theoretical basis for the modern computer by quantifying logical thought. In 1854, his *Investigation into the Laws of Thought* presented a new kind of algebra, in which mathematical equations were represented by logical statements that could take only one of the two values "true" or "false." This Boolean algebra had no immediate application, but its binary nature proved compelling when it became apparent that electricity was the preferred medium for computers: The simplest conceivable electrical device, the on-off switch, controls exactly two states—current flowing or not flowing—which can just as easily be labeled "true" or "false."

As irony would have it, difficulties in mechanical computation like those that stymied Babbage inspired the U.S. mathematician Claude Shannon to apply Boole's ideas. In 1936, Shannon, a graduate student at MIT, was analyzing the behavior of a mechanical computer called the Differential Analyzer—a useful machine, but one that was awkward to program and maintain. Shannon concluded that its operations could be better accomplished with electricity. He had taken a course in Boolean algebra, and in a landmark 1938 paper adapted from his master's thesis, pointed out that a collection of on-off switches arranged according to Boolean principles could carry out logical and mathematical operations. (Shannon was later to say, "It just happened that no one else was familiar with both fields at the same time." He went on to Bell Telephone Laboratories, where in 1948 he wrote another seminal paper, "A Mathematical Theory of Communication" that laid the basis for information theory.)

The only drawback to this scheme was that it forced the computer to operate with a binary number system rather than the familiar decimal one. That was the birth of the *binary digit* or *bit,* which takes on only a value of 0 or 1. Numbers are very long in this system: for instance, the decimal number 31 becomes the binary number 11111. But the advantages of working with a two-state electrical system far outweighed this slight complication, and the computer could always be programmed to deal with input and output in the decimal form favored by humans.

Then it became a matter of engineering to implement Shannon's ideas. The first programmable binary calculator was built in 1938 by Konrad Zuse in Berlin, as a mechanical device to illustrate the principle. This was followed in the 1940s by electric Boolean computers, some of which used electromechanical relays, on-off switches that operate by remote control. An electric current is sent through a coil of wire, producing a magnetic field that pulls a metal finger so that it makes or breaks an electrical circuit.

Relays had been highly developed for telephone networks, which require myriads of choices to route calls, and an early relay-based computer was built at Bell Labs. In 1941, Zuse built an electrical version that worked much faster than a mechanical unit, but in one way, the machine was inferior to Babbage's ideal machine—it could not perform conditional jumps. The ultimate relay-based computer was the Harvard-IBM Automatic Sequence Controlled Calculator ("Mark I"), built at Harvard in 1943. This enormous machine, which calculated gunnery data for the U.S. Navy, weighed 5 tons and was more than 50 feet long. Its miles of wiring linked together more than 500,000 electronic components, including more than 3,000 relays, but for all its massiveness, the Mark I did not support conditional jumps either.

A relay is not the only or best kind of controllable on-off switch. The same function can be performed with a vacuum tube, a device, patented in 1904, that was an outgrowth of Thomas Edison's work with incandescent lighting. A vacuum tube is an evacuated glass envelope that contains electrodes. Without air molecules to interfere, electrons stream through space from electrode to electrode, carrying information and electrical power. This device initiated the electronic age because it could control and amplify electrical signals, making it indispensable for radio and television as well as for video and audio reproduction.

Like a relay, a vacuum tube can switch current on and off, but without mechanical parts, the tube is faster and more reliable than any relay. The first Boolean circuit with tubes was made in 1939, and in 1943, British engineers built the "Colossus." With several thousand vacuum tubes, this special-purpose computer analyzed German military codes as part of the famous "Enigma" code-breaking effort at Bletchley Park in England. The first full-featured electronic digital computer followed in 1946: the electronic numerical integrator and computer (ENIAC) built by J. Presper Eckert and John W. Mauchly at the University of Pennsylvania. Its 18,000 vacuum tubes used many kilowatts of electrical power to determine artillery trajectories at a rate of thousands of calculations a second.

ENIAC served its military purpose and was also used for scientific calculations, but its hardware connections had to be tediously set by hand. Other machines of the era entered programs on punched paper tape and were no great advance either over the Jacquard cards that Babbage had envisioned. The idea that made computers infinitely more flexible is usually ascribed to the brilliant Hungarian-born mathematician John von Neumann, although there is evidence that Eckert, Mauchly and others entertained a smiliar approach. In 1945 von Neumann wrote a report describing the idea of the stored program, where the instructions are held in the computer's memory just as data are. The instructions themselves can be manipulated, making possible, for instance, compilers—programs that convert human-language–like commands into binary-based machine language for the computer. With other features, including a central processing unit and the use of binary numbers and Boolean algebra, this von Neumann architecture is still the standard in computer design. In 1949, the first stored program computer was built by Maurice Wilkes at Cambridge University. Not long after, in 1951, computers came of practical age when Eckert and Mauchly delivered a UNIVAC (Universal Automatic Computer), the first successful commercial electronic computer, to the U.S. Bureau of the Census.

With their size and huge power consumption, these machines were hopeless as candidates for the artificial minds of mobile robots, but the idea of simulating human thinking appeared early in their history. The most significant insights came in 1950 from the British mathematician Alan Turing, whose earlier work had dealt with allied subjects. In 1937, in a paper concerned with the nature of mathematical proof, he proposed a method to break any mathematical problem into a series of steps. This is exactly how a computer program works, and so although Turing was not writing about computers per se, his process amounted to a theoretical description of a modern computer before a single one had been built.

During World War II, Turing, as one of the team of analysts working on Enigma code breaking, had an opportunity to come into contact with real computers. Although much secrecy surrounded the project, it seems likely, as Andrew Hodges notes in his book *Alan Turing: the Enigma,* that Turing was exposed to the capabilities of the Colossus computer. In any event, in 1950, Turing wrote the seminal paper "Computing Machinery and Intelligence," with the opening sentence "I propose to consider the question 'Can machines think?' "

Turing believed that if a computer could do any and all mathematical operations, "We may hope that machines will eventually compete with men in all purely intellectual fields," and proposed the now-famous Turing test as a meaningful measure of machine intelligence. Writing in 1950, Turing stated his belief that

> in about fifty years time it will be possible to programme computers ... so well that an average interrogator will not have more than 70 per cent chance of making the right identification after five minutes of questioning. . . . I believe that at the end of the century . . . general educated opinion will have altered so much that one will be able to speak of machines thinking without expecting to be contradicted.

Turing estimated that a computer with a storage capacity of about 1 billion bits could pass his test. In a way, he was a twentieth-century Babbage, because that requirement was exponentially beyond the technology of the time, as were other ideas of his, for instance, that an important part of machine intelligence would arise by enabling the computer to learn.

Turing was not alone in believing that machine intelligence could be realized, or at least was worth investigating. Six years after his paper, the first study group on the subject was convened at Dartmouth College by the mathematician John McCarthy, who coined the term "artificial intelligence." Other attendees included Claude Shannon and Marvin Minsky, who was to become a highly influential pioneer in the field at MIT. The conference manifesto read

> The study is to proceed on the basis of the conjecture that every aspect of learning or any other feature of intelligence can in principle be so precisely described that a machine can be made to simulate it.

The strategy that emerged was the development of programmed simulations of important chunks of intelligent human behavior. Language skills are one extremely significant part of our thought processes, and early AI researchers worked on machine translation of language, as well as natural language processing; that is, communicating with computers in ordinary language, not special programming languages. Another chunk is the mix of logical thought and strategic planning exemplified in game playing, chess being a prime example. A third is the deductive thinking used in mathematical and geometric proofs. And finally there is visual cognition, the ability to see and give meaning to a scene—among the most challenging of higher brain functions.

While early AI researchers were programming machines to think intelligently in these areas—or at least, trying to—the science of artificial beings was developing in ways that became increasingly entwined with AI and computers. The first more-or-less humanoid creations appeared in the 1920s and 1930s (by then, following Carel Capek's *R.U.R*, such creations were called "robots."). One early model was displayed in London in 1928. It did not walk but could move its arms, hands, and head, rise from a seat and take a bow, and speak by way of a voice box, although what it said is no longer known. It was animated by an electric motor driving an array of cables and pulleys that the early Greeks would have recognized, with electromagnets providing additional flexibility.

A decade later, a more sophisticated example of a robot was extremely popular at the 1939 New York World's Fair, a showplace for the technology that would supposedly improve the world. Elektro the robot was constructed by the Westinghouse corporation. This 8-foot-tall metal construction could move forward and backward, count to 10, and say 77 words. Although Elektro was a large, threatening-looking clanker, Westinghouse went out of its way to humanize the robot. It could dance, and smoke a cigarette, which at the time also seemed endearingly human. A contemporary photograph shows a woman offering Elektro's robot dog Sparko a tidbit as the creature sits up and begs. The woman is tiny compared to Elektro, but the robot stands benevolently by and the whole scene radiates friendly technology.

More than 60 years after that World's Fair, Elektro's engineering details are difficult to come by, but most likely it carried out fixed routines controlled by the relays and vacuum tubes then being introduced into computers. This was the technology Isaac Asimov alluded to in *I, Robot* as inadequate for versatile behavior without a "positronic brain"; relays and tubes alone were not enough to support complex robotic thoughts and actions.

But as computers and AI developed, the "positronic brain" came closer to realization. First, electronic brains had to become smaller and less power-hungry if they were to be installed in robots. The march toward solid-state electronics took care of much of that. Bulky vacuum tubes gave way to tiny transistors (invented in 1947 by

William Shockley, John Bardeen, and Walter Brattain at Bell Labs). These new devices immediately enabled the construction of improved computers, which were soon employed to control so-called industrial robots. George Devol, an engineer, patented the first such device in 1954, and with his partner Joseph Engelberger founded a company to make and sell the UNIMATE—a programmable, one-armed manipulator for use in assembly lines and industrial processes. Engelberger saw such robots as "help[ing] the factory operator in a way that can be compared to business machines as an aid to the office worker."

General Motors bought its first UNIMATE in 1961, but despite Engelberger's optimism, these robots did not become widespread in the U.S. automobile industry until their economic advantages became apparent—especially in competition with Japanese industry, which began enthusiastically adopting industrial robots in the late 1960s. In 1978, GM finally installed a highly automated assembly system that used a programmable arm called PUMA (Programmable Universal Machine for Assembly), and now this type of robot is integral to automobile manufacture and other industries.

Industrial robots are not mobile autonomous mechanisms; they do not move from their bases, and they only follow a preprogrammed series of steps. They are closer to computer-controlled machine tools than to self-determining beings. Nevertheless, they have taught us a great deal about how to make artificial bodies move and how to use computers to control physical actions. The next step was to make smarter artificial minds.

That step was assisted by the advance that came after the invention of solid-state transistors, the invention of integrated circuits in 1958, which put many transistors and other circuit elements on a single tiny piece of silicon. Integrated circuits steadily grew in capacity and shrank in size, going through successive waves—LSI (large-scale integration), VLSI (very large-scale integration), and ULSI (ultralarge-scale integration)—until today a single Pentium-type computer chip contains millions of transistors and other circuit components. These changes reduced the size of computers and powerfully enhanced their speed and storage capacity.

Hopes for successful AI grew along with computer capabilities, but progress was uneven. Programs that produced deductive proofs of logical or mathematical truths, such as those a human mathematician might derive, worked well, perhaps because they were more or less natural extensions of computer processing. But results in machine translation of language were discouraging; available methods could not cope with the subtleties of multiple and contextual meanings of words. And although an AI program could play a good game of checkers, chess was too much. (Even long after these early efforts, in 1997, the chess-playing Deep Blue computer depended on brute force rather than subtle AI-based strategic analysis, using its great speed and memory to examine all possible outcomes of a given move and then selecting the best one.)

In these early approaches, the idea was to put into the computer a complete model of a system in symbolic terms that the computer could incorporate and apply. But it became clear that this "top-down" or "symbolic AI" method was not necessarily the best technique for robots operating in the real world. One famous example of the symbolic approach comes from work on robotic vision carried out from 1968 to 1972 at the Stanford Research Institute (now SRI International). Funded by the Department of Defense, the plan was to make a robot that could autonomously traverse a battlefield to deliver supplies and gather fire-control information. The test unit (dubbed "Shakey" because it wobbled as it moved) consisted of a motorized wheeled platform with a computer, a TV camera for vision, a rangefinder to measure distance to an object, and a radio link to a second, remote, computer for more processing capacity. The robot was developed in an idealized environment, a set of rooms containing simple, brightly colored shapes such as cubes.

Shakey would receive a typed command such as "Find the cube-shaped block in that room and push it to the other room." The robot would examine the room and the objects in it, identify the target, plan a route that avoided obstacles, and carry out the planned moves. Within this laborious process, Shakey displayed flashes of intelligence that combined perception, problem-solving capability, and the ability to move to the right place. In one trial, Shakey shifted a ramp so that the robot could roll up it to reach a target on a raised platform. But the calculations required to accomplish such tasks took hours; even worse, the robot could not cope with changes such as a rearrangement of the objects in its special environment, let alone deal with the infinitely more complex conditions found on a battlefield. Shakey's top-down approach could not pre-factor in every possibility, and it produced an entity far less adaptable than a human or, for that matter, a dog or cat, which knows how to avoid obstacles even in a strange environment.

In the mid-1980s, robotics researcher Rodney Brooks (then at Stanford University, now at MIT) found himself dissatisfied with this kind of limited performance and began questioning the value of the symbolic approach. Speaking of "intelligence without representation," he proposed that robots could act intelligently without using internal symbols. The mobile units he built could be called stupid, in that their programming and computing power were less rich than Shakey's, and instead of the "brain" being localized the processors were distributed throughout the robots to control their individual parts. Further, the sensors that detected how the robot interacted with the real world were closely tied to the motors that controlled its actions, so that the unit could respond rapidly to the data flowing in.

The result was that Brooks's robots evolved their own behavior as they explored the world. For example, one of his early efforts, called Genghis, learned to walk. Although wheeled robots have their uses, a robot with legs manages better in rough terrain, which might be encountered when NASA sends robotic explorers to distant planets. Insectlike, Genghis had six legs, each with its own motors, processor, and sensors that registered what the leg was doing. Additional sensors detected obstacles in the robot's path. Others reacted to heat, enabling Genghis to sense the presence of warm-blooded mammals; for example, people.

Initially Genghis's six legs were uncoordinated and the robot could not walk. But as each leg tried different movements, Genghis learned from its mistakes through a form of behavior modification by positive and negative reinforcement. In 1990, Brooks described how the unit was programmed:

> [E]ach of the [robot's] behaviors independently tries to find out (i) whether it is relevant (i. e. whether it is at all correlated to positive feedback) and (ii) what the conditions are under which it becomes reliable (i.e. the conditions under which it maximizes the probability of receiving positive feedback and minimizes the probability of receiving negative feedback).

"Positive" and "negative" feedback mean that the signals to the leg motors are modified to enhance or diminish the occurrence of specific motions, depending on whether they contribute to the goal of walking—a process similar in spirit to biological evolution, which by trial and error weeds out whatever does not contribute to an organism's survival. Little by little, the six legs coordinated themselves and Genghis became a sophisticated walker. The result is a robot that behaves in a lifelike manner as it crawls on the floor and over obstacles, and follows a human around the room when its heat sensors detect one.

In a top-down approach, a robot's actions are motivated by the expectations that are part of the symbolic model of the world that is built into it; but Genghis gained the intelligence to walk by responding directly to stimuli, an approach often called bottom-up. Both the top-down and bottom-up methods are valuable in constructing artificial minds even if they lack bodies, but it is easy to see that the latter approach is especially meaningful for a robot that physically interacts with the real world. An autonomous robot is not useful unless it deals intelligently with its physical environment, where it has to move without collisions, manipulate objects without breaking them, and so on. If the right learning mechanisms could be found, that interaction would constantly help the robotic brain develop on the basis of experience, just as we humans learn to function in the world by doing things in it.

Brooks's approach was one new thread in AI that began in the mid-1980s; it was not the only one. In 1986, Marvin Minsky presented an alternate approach in his book *The Society of Mind*. Rather than consider the human mind as a single entity responsible for all thought and behavior, which could in principle be described once and for all, he proposed that different components of the brain all "speak" at the same time. From this babble, in which some voices are loud and others soft, some agree and some oppose each other, a consensus emerges that defines behavior.

The idea of distributed and even contending "voices" within the human mind might seem strange. Minsky's model is only one among many proposed since the time of Descartes, as we consider how the physical brain and conscious mind are linked to each other, and is far from being accepted as a definitive explanation of how the mind works. But used as an AI technique, the concept of multiple voices has imparted convincingly lifelike behavior to various robotic toys.

Along with such changes in the design of artificial brains, enhancements in computer speed and capacity offered new possibilities for AI. One advance was the technique of parallel processing, which some observers considered a fifth generation in computing (the fourth generation consisting of computers using VLSI and ULSI technology). In a parallel processor, many computer chips are interconnected so that each one handles a different part of a problem at the same time, which can give impressive results. In 1987, for instance, a parallel processor called the Connection Machine operated 64,000 microprocessors simultaneously to perform two billion computer operations per second—an impressively high speed that could hardly be matched by conventional computers at the time. But programming a parallel machine so that the parts of the problem are properly parceled out is difficult, and it is unclear whether parallel processing can offer enormous advantages.

However, the idea of carrying out many "thought" operations at the same time is promising for AI because that's how the human brain works. Each of its many billions of neurons is intricately connected with others through upward of a thousand connecting points, called synapses. The neural signals that define the brain's operations travel through the network. Many neural events are going on at the same time, a huge benefit for processing speed. The multiply connected neural architecture also protects a brain that is partly damaged from necessarily losing an entire function such as memory, and allows replacement neural connections to be forged so that new areas can take over from damaged ones. In fact, the process of learning seems to consist of recording the new knowledge by means of new connections that form among neurons. This property of the brain is known as plasticity.

The highly connected architecture of the brain is a model for another approach to AI, the neural net. Unlike a conventional computer, such a net more or less simulates real biological brains. Analogous to the web of neurons that makes up a natural brain, a neural net consists of many simple processing units interconnected so that they can trade data, with each unit operating on the data it receives. Depending on how the net is structured, the system can acquire and store knowledge through a learning process that might teach it, for instance, how to identify particular images or sounds.

In 1943, Warren McCulloch and Walter Pitts, at the University of Illinois, laid the groundwork for neural nets through pioneering research that depended on viewing the brain as a complex network of neural elements. In order to make a reading machine for the blind that turned printed material into sounded-out words, they interconnected light detectors in a way that mimicked neural connections in the brain. In 1951, Marvin Minsky and a collaborator constructed another neural net called the Stochastic Neural-Analog Reinforcement Computer (SNARC), which was trained to negotiate a maze as a rat would. Further work in neural nets focused on recognition of visual and aural patterns, but the approach languished in 1968 when Minsky and his colleague Seymour Papert pointed out limitations in the method as then understood. New insights, however, revived the technique in the mid-1980s, along with other approaches that approximate biological styles of thinking.

SENSING

Initially, AI researchers aimed to produce intelligence within a computer, not a robot. A computer interacts with other machines or humans through the abstract medium of data flow but has no direct connection to its physical environment. An operational robot is different; it must take in information from its surroundings and respond in real time. It is meant to emulate what happens in a human, in whom the five senses gather information about the exterior world that is sent to the brain for analysis and response. An artificial brain in a blind and deaf robot body would be useless; it needs sensors that simulate the human ones, or go beyond them.

As electronic science and technology developed, they led to the construction of artificial sensors analogous to the human senses. Machine vision had roots in late-nineteenth-century discoveries that light could change the electrical properties of certain materials or cause them to emit a flow of electrons; that is, an electrical current (one of these phenomena, the photoelectric effect, so baffled physicists that Albert Einstein earned a Nobel Prize in 1921 for explaining it). These effects were incorporated into devices that detected light by turning it into electricity. By the 1920s, the conversion of light into electrical signals was advanced enough that television images were being broadcast on an experimental basis, and in the late 1930s, the BBC in England and the RCA Corporation in the United States began regular television broadcasting. Improved television cameras were developed in 1939. Following World War II, commercial television broadcasting became widespread, and in 1953, color television was introduced.

Within the digital revolution brought on by the growth of computation, the development of video cameras whose electrical data came as a stream of binary digits was inevitable. Like a human eye that gathers data and sends it to the brain for analysis, a computer-based camera scans a scene and presents it to the computer for further processing—but this is only the beginning of meaningful machine vision. The quantity of data involved in a pictorial representation of the world is staggering and requires extremely high levels of computational power to process. And as the Shakey robot demonstrated, gathering visual data and transmitting them to the computer is the easy part; it is extraordinarily difficult to decide how to assign meaning to an image so that the robot can act on the information.

Similarly, artificial sensing of sound and means to generate it began in the late nineteenth century. Between 1876 and 1878, Alexander Graham Bell patented the telephone, Thomas Edison patented the phonograph, and the British musician David Edward Hughes invented a microphone in which the pressure from sound waves altered the electrical properties of grains of carbon. Later innovations included magnetic recording, first proposed in 1920, and high-quality sound reproduction along with stereophonic sound, which began in the late 1950s and early 1960s.

Even early in the history of computers, it was music that drove the further development of digital sound techniques. The first music synthesizer program was written at Bell Labs in 1960, and by 1984, a set of standards had been created to transmit musical information in digital form between electronic synthesizers and computers. Now digital recording and playback of music, and word recognition and synthesis, are routine functions even on small desktop computers. But as with vision, the mere ability to register or produce sounds or words under computer control does not give them meaning. In humans, the linguistic analysis that the brain performs as we speak and listen is one of our most demanding intellectual functions.

Whereas the mechanics of artificial vision and hearing have been refined by decades of development, synthetic touch, taste, and smell are less highly evolved, partly because their nature in humans and animals is not so well understood. Touch and taste have the complication of operating over large areas with many sensors, taste and smell involve varied chemical interactions, and taste also seems to depend on texture. Still, artificial versions of all these senses exist and are being steadily improved. Touch has been implemented by sensors that produce an electrical effect when they are deflected or change position in space; these serve as collision-avoiding devices and enable an artificial being to judge its bodily orientation. Finer tactile sensing, like that of the human fingertips, is also under development, as are analogues for smell and taste.

In practice, these last two senses are probably the least essential for an artificial being, which could be highly functional with only vision, hearing, and a limited sense of touch. But the sense of smell carries a special meaning for us and illustrates the complexities of simulating human behavior. Although smell does not play the central role for humans that it does for many other animals, it is in a way our most fundamental sense. It connects directly to an ancient part of the brain, the limbic system, without the high-level processing that vision entails. That is why an odor can evoke mood or feeling instantly. As we'll see later, emotional factors like these might be surprisingly relevant to the creation of intelligent artificial beings.

If there are lacks in artificial senses, there are also compensations, because the natural senses have limitations. The wavelengths of light we see are a small fraction of the range of electromagnetic wavelengths in the universe, which includes infrared, ultraviolet, and more. But there are sensors that detect radiation at these wavelengths and give artificial vision extrahuman capabilities; infrared vision, for instance, can penetrate darkness. Other possibilities abound, such as using sonar the way submarines do to probe the environment, as well as the functional equivalent of telepathy—direct mind-to-mind communication among artificial beings by radio.

The possibilities for touch, smell, and taste, and for extrahuman senses, are new enough that their further discussion belongs in the second half of this book. But another aspect of artificial creatures, their appearance, has roots that go back to mechanical automata.

LOOKING HUMAN

For all their developing mental, sensory, and physical capacities, modern digital artificial beings are inferior in one way to the eighteenth-century automata of Jaquet-Droz and de Vaucanson; they are not androids, they do not look human. The artisans who built clockwork automata took great pains to make their creations resemble people, modeling the faces and hair, dressing them well, and aiding the illusion with subtle but telling cues to humanness, such as having Henri Maillardet's "Draughtsman-Writer" look down at the paper before starting to write.

Most modern artificial beings, however, do not look like real people. The 1939 World's Fair clanker robot Elektro had a humanoid outline, with limbs, torso, and a head, but its size, metal body, and cartoonish facial features unquestionably connoted a machine. More recent robots like Shakey, Genghis, and Cog are even less prepossessing as human stand-ins, and indeed were not created with that aim in mind. As special-purpose or test units, there was no advantage in making them humanlike. They are instead bare assemblages of wheels or legs, motors, girders, sensors and computer processors arranged for engineering convenience or to provide specific physical abilities.

In the 80 years that electrical and electronic robots have existed, we have yet to create an autonomous, human-seeming android like those that appear in the virtual history of artificial beings. But the introduction of robotics technology to the entertainment industry has brought us some way toward natural-looking artificial creatures, in the development of so-called animatronic figures. The Walt Disney Corporation introduced these three-dimensional entertainment robots in human and animal form at Disneyland in 1963, and showed them at the New York World's Fair in 1964. The first was a tap-dancing simulacrum of the dancer and actor Buddy Ebsen. Others included Abraham Lincoln standing, speaking, and gesturing, a dinosaur diorama, and the exhibit, *The World of Tomorrow.*

The original animatronic robots, and current versions that appear in films, are not autonomous. They operate under remote control from human operators or, like an industrial robot, perform an unvarying computer-controlled sequence of actions. But they show how far we have come in producing artificial beings that look convincingly natural. Their development has required new styles of engineering to avoid making awkward clankers. Nick Maley, who has worked on varied animatronic "creature effects" including the character Yoda in the 1980 film *The Empire Strikes Back* notes subtle differences between standard methods and what is effective in replicating living beings. Engineers constructing mechanical beings, he says, tend to use

> strong materials to build robust mechanisms based upon the same tried and tested mechanical principles that create cars and trains. . . . However, nature's creations don't use the same mechanical principles. . . . Their joints are less precise, their connections less rigid. . . . Their existence is usually a delicate balance of strength and weight developed to suit specific circumstances.

Maintaining that delicate balance is not easy. One recent animatronic model of a human head is packed to the very skin with a dense collection of components and wires, including 22 small motors that control its movements and facial features.

Along with flexible mechatronic design, these robots use materials that replace the substances once incorporated into eighteenth-century automata. Glass eyes were first made in Venice around 1579, but have been replaced by plastic versions that look more natural. Likewise, although prosthetic limbs, from wooden legs to iron hands, have a long history, they began to look convincing only with the arrival of silicone rubber, a compound with natural-feeling resilient properties. It can be used to form an artificial skin with layers that approximate the internal structure, and therefore the feel, of real skin, it can be colored as desired, and pores and hair can be added as final persuasive details.

These and other advances are leading simultaneously to improved prosthetic devices and to the possibility of androids whose internal structure is overlaid with an artificial humanlike outer layer—modern versions of eighteenth-century automata. Like the rubber in de Vauconson's duck that simulated the digestive tract, some of the internal machinery functionally replicates what goes on in living creatures; for instance, by using "smart materials," which change their properties under external control. One type can be made to extend and contract depending on electrical voltage, simulating how human muscles act. The result is a synthetic muscle that can give artificial limbs a smooth, natural action, rather than the jerkier motion produced by machinery.

Such humanlike flexibility opens up possibilities for convincing bodily motion and even facial expression. For instance, the Saya robot at the Tokyo University of Science has a humanlike face with silicone-rubber skin. Underlying this is a set of artificial muscles, worked by compressed air and arranged to follow human facial anatomy, that can be manipulated to display joy, anger, astonishment, and other emotions. Although the question whether artificial beings can or should experience emotions is complex, there is no doubt that the ability to simulate emotion through facial expression and body language greatly affects interactions with people.

Artificial skin and muscles are examples of how the technology of artificial creatures and research in human implants interact with each other to create replacements for body parts and organs and, potentially, improvements in human function. The medical market for implanted devices is enormous, with millions of implants performed every year. This biomedical enterprise provides a technological base for efforts to enable artificial beings to mimic human capabilities, while research in artificial beings leads to better implants.

Despite our best efforts to construct artificial beings, at the moment living organs, developed through millennia of evolutionary progress, are generally superior to their artificial counterparts. Even a cat or mouse brain, let alone a human one, functions more intelligently in the real world than the best AI-driven robot yet built. The sensitive nose of a dog detects odors beyond the capabilities of mechanical sniffers. The answer to some of the pressing problems of creating artificial creatures might be the combining of nonliving with living parts, just as the god Hephaestus used the flow of ichor—blood—to add something essential to his bronze robot, Talos. As the next chapter shows, humanity already has a surprising history of combining the living and the artificial.

4

We Have Always Been Bionic

Among the most intriguing beings in the virtual history are those that combine the living with the nonliving, such as the cyborgs Deirdre and RoboCop. These particular examples consist of a machinelike body of superior physical capability that is controlled by an implanted human brain. A hybrid being might also begin as an ordinary human, who is significantly modified with artificial parts or implants. (This is how the Tin Woodman became what he is in *The Wizard of Oz:* he started as a human, but as he accidentally chopped bits off himself and had them replaced by a tinsmith, he eventually became wholly metallic.) In either case, there is poignancy in the merging of human softness and frailty with the hard precision and power of a machine, and in the extreme, in the image of a mind and spirit isolated from the run of humanity within a dead shell.

It is easy to imagine such a hybrid as a spiritual amphibian, infinitely more displaced and alienated than, say, a person caught between two cultures and not fully belonging to either. At a deeper level, human–machine amphibians force us into a close examination of what living and nonliving really mean. In these beings, the boundary between the two states blurs, providing a third mode of existence that lies somewhere between unfeeling machine and feeling human.

However, in the real rather than the virtual world, there are as yet no human brains operating in artificial bodies. And although bionic people have been around for a long time, until recently their artificial parts have been primarily mechanical, not neural, and represent relatively small bodily changes. Replacements for missing limbs and cosmetic additions such as breast implants are immensely significant to the implantee, but they do not turn people into cyborgs—nor is anyone yet proposing to transplant a living brain into a metal body. Still, the latest chapter in the real history of artificial beings is a step toward this fusion; it is the formation of direct connections between living organic systems and nonliving ones at the neural and brain levels.

The key idea behind this synthesis draws on the electrical nature of the signals in the nerve network and the brain and envisages connecting neural systems to electronic ones. Outcomes already beginning to be realized include an interface that allows a paralyzed person to manipulate a computer purely by mental control, without physical effort; hybrid neural-electronic chips, in which a living neuron and an electronic circuit mounted on the same piece of silicon communicate with each other; and the use of animal brains to control mobile bodies and robotic arms, as a step toward providing mentally controlled devices to the paralyzed.

These developments fall under new areas in neural research and clinical practice called neurorobotics or neuroprosthesis, but artificial additions to human bodies have a long history that—like the development of artificial beings themselves—reflects successive waves of technology. Prostheses that use digital electronics owe their invention, in turn, to the development of implant surgery and to the scientific introduction of electricity into the body. But first came simpler physical prostheses without electronic components, made to meet the needs of those born handicapped, injured through accident, or wounded in war. Such prostheses fell into two categories: functional, to replace lost physical capability, and cosmetic, to rehabilitate damaged appearance.

UNFEELING LIMBS

Both functional and cosmetic rehabilitation were recognized in the early virtual history of bionic beings. What is said to be the first prosthetic described in writing appeared two to three millennia BCE in the Indian Rig-Veda poem, in which Queen Vishpla, having lost a leg in battle, replaces it with an iron one and returns to the fight. Other prosthetic devices have appeared in Greek mythology. Although Hephaestus, that limping Greek god of technology, did not use an artificial limb, he relied on a crutch and on the help of the golden assistants he had constructed. In an especially gruesome Greek tale, Tantalus, son of Zeus, killed and cooked his son Pelops and served him to the gods to see if they could distinguish between human and animal flesh. After Demeter, goddess of agriculture, ate Pelops's shoulder, she atoned by restoring him to life complete with a new ivory shoulder. Prosthetics entered Greek culture in a different way in the fifth century BCE, when Aristophanes' play *The Birds* included a character with a wooden leg.

Wooden replacements for legs and feet are among the earliest examples of real, as distinct from imaginary, prosthetic devices. About 440 BCE, the Greek historian Herodotus wrote of the Persian Hegistratus, who was captured by the Spartans and held captive by having his leg locked into a wooden stock. To escape, he amputated part of his foot so that he could pull it through the hole, and later replaced the missing part with a wooden substitute. The Romans also constructed replacements for missing hands, and by medieval times, wooden peg legs or iron hooks had become the standard replacements for missing legs or hands.

There was nothing aesthetically pleasing about peg legs, but they were a simple way to support body weight. The same can be said of a hook in lieu of a hand; it gives limited ability to manipulate objects without matching either the look or the usefulness of a true hand. It was difficult for ancient artificers to make prostheses that both looked the part and acted it, but sometimes, cosmetic appearance and proper functionality could be combined. One of the older prostheses found by archeologists is Roman and dates to about 300 BCE. Made of bronze and wood, it is modeled to resemble a leg from thigh to calf. Some prosthetic devices, however, were purely cosmetic, such as the metal nose (supposedly made of alloyed gold, silver, and perhaps copper) with which the sixteenth-century Danish astronomer Tycho Brahe replaced his real nose after it was sliced off in a duel. Under the same heading come cosmetic additions and replacements still used today: hair implants for men, breast implants for women, and nonfunctional glass or plastic eyes for both.

Purely cosmetic replacements are widespread bionic additions that are deeply important to their users, but they offer a lesser challenge than functional prosthetic devices that replicate human abilities. Much of the impetus to make artificial limbs that actually work has come from the needs of injured warriors and soldiers. The knights of medieval Europe in particular had a certain advantage: Their metal armor required the services of armorers, and these artisans were also capable of designing and making functional devices to replace limbs lost in battle. Because knights in armor were already clad in metal, the replacements matched the missing limb in appearance so they worked cosmetically as well.

Some of these knightly prosthetics showed truly advanced features. The most famous example was fashioned for the German knight Götz von Berlichingen, also called Götz mit der Eisernen Hand; that is, Götz with the Iron Hand. Known as a kind of Robin Hood who took the side of peasants against their oppressors, his story was told in the play named after him, written by Johann Wolfgang von Goethe.

Von Berlichingen lost his right hand from a cannon-ball strike at the battle of Landshut in 1504. He had it replaced with an iron prosthesis that featured movable fingers that could be adjusted by his natural hand and locked into place or released through an arrangement of springs. The entire artificial hand could also be set into varied positions. This was not even the first or only such adjustable hand; another with similar characteristics, found near the river Rhine, is thought to date to 1400. A later iron hand and arm, dated about 1602, would look perfectly at home attached to a modern clanker robot. Other medieval prosthetics were designed for specialized knightly needs, such as an artificial knee built in a semiflexed position that allowed a knight to ride his steed, although it did not support sitting or standing.

Sixteenth-century warfare also motivated the French physician Ambroise Paré to develop innovative procedures that made him a founding figure for modern surgical practice and amputation medicine. His wide experience as an army surgeon gave him ample acquaintance with severe injuries, and he introduced artificial eyes (made of gold and silver) and teeth, and a prosthetic leg. One invention, "Le Petit Lorrain," was a hand operated by springs that an officer in the French army used in battle.

As in the history of automata, this phase of the development of prosthetics relied on the work of mechanical experts such as armorers and watchmakers, and on the growing knowledge of anatomy. But still the technology was not sufficiently advanced to make devices that were both functional and natural looking, or to make limbs that were easy to use. Iron prosthetics were heavy, and their only source of power was either a natural hand that set and adjusted the artificial unit, or other muscles in the body. Beginning in 1818 and continuing to modern times, inventors have developed harnesses and levers that carry power from other parts of the body, such as the shoulder, to make an artificial hand, say, open and close its grasp.

Natural appearance often had to be sacrificed to functionality, and power to operate a limb was hard to come by. Nevertheless, early inventors improved prosthetic devices through the ingenious use of materials. In 1800, for instance, James Potts of London designed a false limb that came to be known as the "Anglesey Leg," because it replaced a leg lost by the Marquis of Anglesey at the Battle of Waterloo. Among its advanced features was an articulated foot that could be controlled by catgut strings, extending from knee to ankle, which determined the position of the foot by transmitting motion from the knee. These cablelike control elements have natural parallels; for example, tendons that stretch back to muscles in the arm control the fingers of our hands. Along similar lines, one modern breakthrough is the development of artificial muscles that work like real ones.

More than a century later, around 1912, the English aviator Marcel Desoutter began a trend toward lightness and durability with the introduction of aluminum as a prosthetic material. Although pure aluminum was first extracted in 1827, it was so expensive to produce that it was used mostly in jewelry throughout much of the nineteenth century. But after a cheaper manufacturing method was invented in 1886, aluminum entered industrial use. Its use in aircraft began in 1897 when it was used to form the frame of an airship, and it continued to play an important role in aviation. When Desoutter lost a leg in an airplane accident, he and his brother, an aeronautical engineer, designed the first prosthesis to use aluminum, combining strength with lightness.

While the needs of knightly warriors had provided initial motivation, and advances came from individual efforts like those of Marcel Desoutter, it took the massive scale of modern warfare to truly stimulate prosthetic science. In the American Civil War, the combination of enormous casualties with the state of nineteenth-century medical practice meant that amputations were common—30,000 on the Union side alone. (On the Confederate side, General John Hood had his right leg amputated after he was shot at the battle of Chickamauga in 1863. He finished out the war with a wooden leg that allowed him to continue riding horseback.) When, in 1862, the federal government guaranteed prostheses for Union veterans who had lost limbs, the result was the growth of a business that by 1917 supported some 200 clinics. World War I also had its effect, albeit a relatively limited one in the United States, which was involved in the war only from 1917 to 1918. American soldiers suffered more than 4,000 amputations, compared to nearly 10 times as many for British troops and a total of 100,000 for all the armies from European nations—a number that inspired the growth of prosthetic technology in Europe.

But after World War II, with its extensive casualties among all the combatants (including more than 45,000 amputees among U.S. troops), the need for the serious development of prostheses became widely recognized. Improvements proceeded faster, encouraged by government support. Now, although we do not have a major conflict on the scale of a world war, prosthetics are still needed to replace limbs lost or amputated through accident and disease. The U.S. population includes more than one million amputees, with an estimated 100,000 lower-limb amputees added yearly. And in some parts of the world, there is a residue of war that maims thousands of people a year—the unexploded land mines strewn around many countries, from Afghanistan to Mozambique. Land mines are cheap and effective weapons, and estimates range up to 100 million of them buried in 62 countries, with Cambodia having one of the densest concentrations. The result is that the business of making prostheses, along with the allied industry of orthotics (limb braces), is estimated to be a $2 billion undertaking worldwide.

This industry has seen significant technological progress. Where metals are used, they are the lightest available, including titanium, but increasingly they are replaced by new materials such as graphite composites like those used in tennis rackets, and plastics, which can be formed into natural-appearing limbs. The mechanical systems that articulate the limbs have also been improved, using pneumatic or hydraulic fittings to provide smooth motion. Some artificial legs are good enough that their wearers can enter athletic events with satisfying performances, such as a time of 12.4 seconds for the 100-meter dash turned in by one runner equipped with a prosthetic leg.

The power sources that move artificial limbs have become more sophisticated as well. Energy-storing artificial feet, designed in the 1980s, incorporate a spring that compresses as the foot strikes the ground, and then extends to release the stored energy and help propel the leg into the next stride forward. Extremely small electric motors have also been developed. Some of them are tiny enough to fit into artificial fingers and hands and powerful enough, for instance, to provide a grasping function, while drawing so little electrical energy that battery operation seems feasible.

However, no matter how effective the engineering and aesthetic design of an artificial arm or leg, it still lacks an important capability. An artificial leg has no sensors to test the nature of the walking surface in order to adjust pace and maintain balance, nor does it receive commands from a brain that brings in other sensory information such as visual data to forecast changing circumstances. An artificial hand has no sensory feedback that allows the brain to adjust the hand so that it can delicately grasp a teacup without breaking it, or apply full power to twist the cap off a jar. A truly bionic limb needs sensory capability and processing power (in the limb itself, or through connections to the brain) as well as appropriate movement, flexibility, and appearance. In fact, what is needed to make a functional bionic limb for a person is nearly identical to what is needed to make a robotic limb.

This is where digital electronics connects with prosthetic science. Some prosthetic limbs now incorporate electronic sensors and computer chips to make a "smart leg." A direct neural connection between artificial limb and brain is further off, but here, too, initial results have been obtained, using connected or implanted digital electronics. In a way, the eighteenth-century belief that electricity could invigorate the body and even animate a dead one is becoming realizable in the twenty-first century, through artificial devices that operate electronically—a trend that began when electricity was first introduced into the body.

CHARGING THE BODY

In the eighteenth century, although electricity was known to stimulate the body, its physiological effects were not studied in detail. Nineteenth-century physicians began experimenting with the influence of electricity on the heart. In 1888, it was found that fibrillation of the heart—that is, the sudden change of a regular beating pattern into an irregular rapid one—could cause sudden death, and in 1899 researchers found that a strong electrical shock could defibrillate an animal's chaotically beating heart. The first human heart was successfully defibrillated in 1947, and both external and implanted defibrillators were developed in the 1950s and 1960s.

The heart pacemaker is a related and even more helpful bioelectronic device. It sends timed electrical impulses to the heart muscle, setting a proper rate of beating for people whose hearts beat too slowly. Pacemakers are the most widely used functional bionic implants; some two million have been placed into patients around the world since 1960, including one implanted in U.S. vice president Dick Cheney in 2001. The earliest recorded portable models were developed and demonstrated around 1930, but serious development began only in the 1950s. The history of these devices exemplifies many of the general issues associated with bionic implants, along with their benefits.

To provide the stimulating electrical current, some early pacemaker models used electrodes that lay on the skin but did not enter the body, but these were unsuited for long-term use because they burned the skin after a few days. Other versions developed in the 1950s used implanted wires; that is, the electronic part of the pacemaker was mounted outside the body, from where it sent pulses to a small wire that entered the body and made its way to the heart. But this arrangement, too, had serious drawbacks. It was easy for infection to develop at the entry points of the wires. The external electronic unit that generated the proper pulses was too bulky for easy portability and required so much power that it had to draw on conventional house current. This meant that the implantees' mobility was limited by the length of power cords, and the implantees were utterly at the mercy of power failures.

These deficiencies have been remedied through parallel advances in electronics and implantation procedures. The problem of infection could be avoided by implanting the entire unit in the body, but that wasn't possible until the introduction of transistors, which made the units much smaller. As a bonus, the transistorized units also drew less power than the earlier models, so that battery operation became practical. The result was the first wearable battery-powered pacemaker, developed in 1957, and then the first fully implantable unit. The first successful implantation of a pacemaker, in which the unit operated in the implantee for nine months, was carried out in 1960.

Today, further advances in electronics, computation, and implant surgery support highly sophisticated pacemakers. The devices became programmable in the 1970s; that is, their pulse rates could be externally altered by radio signals without additional surgery. Recent models are rate responsive, meaning they detect the implantee's activity and adjust the pulse rate accordingly; they work at minuscule power levels, giving them extremely long lifetimes; they operate in a dual-chamber mode, meaning they use two electrical wires to pace both the upper and lower chambers of the heart, synchronizing blood flow for maximum efficiency; and they store the implantee's medical information in computer memory for retrieval by a physician.

ELECTRIFYING THE MIND

At their high level of perfection, heart pacemakers represent a successful bionic intervention, but they do not involve neural connections. What might be called neurobionics, however, also has a long history arising from the desire to use electricity to affect neural behavior or alleviate certain disabilities. In the Roman era, Scribonius Largus, court physician to the emperor Claudius, reported that he could relieve the pain of headaches by placing a torpedo fish or electric ray—another fish that emits an electric charge—on the sufferer's forehead. Apparently, just as the fish's electric charge stunned its prey, the electricity stunned the patient's nervous system to provide relief. Today electrical stimulation of the nervous system is routinely carried out using both external and implanted devices to relieve pain, and for other therapeutic purposes.

One form of electrical brain stimulation, electroshock or electroconvulsive therapy (ECT), is intended to cure mental disease. The method was conceived when it was seen that epileptic seizures seemed to relieve the symptoms of schizophrenia. By the late 1930s, the Italian researchers Ugo Cerletti and Lucino Bini were learning how to induce such seizures electrically. In initial testing, they placed electrodes so as to send electricity through the entire body of a dog, but the shock to the animal's heart proved fatal. Placing the electrodes on a dog's head, however, avoided any flow of current through the heart. In 1938, electroshock was first applied to a schizophrenic person, who was apparently cured by the procedure, at least for a time. ECT came into heavy use in the 1940s and 1950s, but fell out of favor because of the violent physical convulsions it induced, along with reports of undesirable mental side effects and the possibility of misuse, as dramatically illustrated in the 1975 film *One Flew Over the Cuckoo's Nest.* The introduction of alternatives such as psychiatric drug therapy also made the method less desirable. Recently ECT has seen a comeback in treating severe depression, but the method remains controversial.

Electricity can affect the nerves and the brain in subtle and apparently benign ways as well as in overt and violent ones. Electrical stimulation of the vagus nerve, for example, has reduced the frequency of epileptic effects in many patients, although the reasons for this outcome are not entirely understood. The vagus is a complicated, widely distributed nerve (its name comes from a Latin root meaning "wandering") that runs from the brain stem—which connects the brain to the spinal cord—through the neck and thorax to the abdomen. It has functions related to the ears, tongue, larynx, stomach, and heart. Epilepsy is a chronic disorder of the nervous system, in which seizures arise from excessive interaction among the neurons in the brain. While drugs can reduce that abnormal activity, another possible therapy arose from work dating back to the 1930s, which showed that stimulation of the vagus nerve affects brain activity. In the 1980s, researchers proposed that controlled electrical stimulation of the vagus nerve could desynchronize the brain's neural signals and hence potentially blunt epileptic effects.

That led to the technique called VNS, vagus nerve stimulation, which has proven beneficial for epileptics whose condition is inoperable and does not respond to drugs. In VNS, an electronic pulser the size of a large coin is implanted under the skin on the left side of the patient's chest. Every few minutes, the device—powered by a battery with a lifetime of up to five years—generates a series of electrical pulses that lasts a few seconds. The pulses, typically a few thousandths of an ampere, are sent through a wire wrapped around the portion of the vagus nerve running along the left side of the patient's neck. Patients can also manually activate the device, using a switch operated by an external magnet, when they feel a seizure coming on. The results have been beneficial; studies show that a year after the device is implanted, nearly a quarter of patients have had their seizure rate reduced 90 percent or more.

A similar implanted device is used to alleviate symptoms of Parkinson's disease, a chronic and progressive disorder first called the "Shaking Palsy" in 1817 by the English physician James Parkinson. The disease kills certain neurons in the brain that normally produce the chemical dopamine, which transmits nerve signals among areas in the brain that control the muscles. The nerve damage affects body movements at mild to severe levels, with such symptoms as rigid muscles; tremors of the hands, arms, feet, or jaw; changes in speech and handwriting, and the inability to maintain balance. The symptoms can be treated with drugs that replace the missing dopamine, although they do not halt the neural degeneration. A new approach to relieving the muscular symptoms uses a battery-powered implant, which, like the VNS device, generates electrical pulses, although in this case they are sent deep into a particular region of the brain. Originally approved in 1997 by the Food and Drug Administration (FDA), for implantation on one side of the brain to control tremors on that side of the body, FDA approval was extended in 2002 to allow the implantation of dual systems that operate on both sides of the brain.

The electrical pulses used in VNS or the Parkinson's implant are not digitally encoded, but a more sophisticated type of neural implant does use digital methods to correct another human problem, hearing loss. The physical understanding of sound extends to ancient Greece, where it was realized that sound consists of vibrations in the air. Later, the physiological mechanisms of hearing were explored, illuminating how those vibrations are detected and transmitted in the body. In humans, hearing occurs when sound waves enter the ear canal and set the eardrum vibrating in step with the waves. Those vibrations are transmitted through bony structures to an inner structure called the cochlea. There, the mechanical motion is converted into impulses that travel along the auditory nerves to the brain, where they are analyzed and interpreted to give them meaning.

Earlier methods to improve deficient hearing dealt only with the outer ear. The first approach was the hearing aid, which in early days took the form of an ear horn, a trumpet-shaped device held up to the ear. The ear horn worked like a megaphone or the large hornlike devices seen on Edison's early phonographs, but in reverse; the large cross-sectional area of the horn captured more sound energy than the ear's small opening could and funneled that enhanced sound into the ear itself. With the advent of electricity, however, a hearing aid became something different. It changed sound into an electrical signal that was processed and amplified, and then changed back into a louder, clearer sound fed directly into the hearing-impaired ear through a speaker—but still not going directly into the auditory nerves.

Electrical hearing aids were in use by the late nineteenth century. An electrical unit called the Akoulathon, invented in 1898, was being sold commercially in 1901. Like early telephones, it used a carbon "transmitter" or microphone of the type invented in the 1870s. Then, as in the development of computers, advances in electronics—first the vacuum tube that amplified the signals going to the ear, followed by transistors and integrated circuits—led to today's extremely small and efficient hearing aids.

DIGITAL EARS

Today's hearing aids certainly help those with hearing loss but do not qualify as true bionic additions because they are not directly implanted into the body or connected to its neural system. But help for the hearing-impaired reached bionic status in the late 1950s, when several researchers explored the possibility of direct electrical stimulation of the cochlear nerves. The expectation was that if the nerves were intact, stimulating them directly might produce the sensation of sound in the brain. Considerable development led to the cochlear implant, today's most mature neural prosthesis—the only one that is commercially available—and the most widespread, with some 30,000 implanted since 1999.

The cochlea—named after the Latin word for "snail"—is a

hollow, fluid-filled structure shaped like a snail shell that resides in the inner ear. Uncoiled, it would stretch well over an inch, but in its natural state it is the size of a pea. Its nearly three full turns contain the nerve endings that make human hearing possible. The process begins when the sound vibrations detected by the eardrum enter the cochlea, where they set internal structures into corresponding vibration. This in turn affects bundles of hairs growing out of sensing units called hair cells. Through a complex mechanical and electrochemical process, the motion of the hairs is converted into electrical signals that travel through the cochlear nerves to the auditory cortex, the part of the brain that interprets the signals as sound.

To perform the artificial equivalent of this natural process, a cochlear implant is surgically embedded in the skull just behind the ear. An external microphone worn behind the ear picks up sound and sends it to a processor, also external. The processor amplifies the sound, filters out extraneous noises, and converts the result into digital electronic impulses that go to a wireless transmitter worn behind the ear, which sends the pulses to a receiver implanted under the skin. The receiver picks up the signals and sends them along wires—up to 24 of them—bundled into a narrow tube that has been woven into the cochlea. There, the digital signals stimulate the auditory nerves to produce neural impulses that are interpreted by the brain as specific sounds. The entire affair is operated by a small battery.

The cochlear implant restores a greater or lesser level of hearing in many deaf implantees. While the sensitivity of the device is too low to allow the listener to hear the very softest sounds, medium- to high-level sounds can be heard. Almost one-third of cochlear implantees hear spoken words clearly enough to use the telephone, and about half of implanted adults who knew how to speak before they lost their hearing can understand at least some words. Even those who do not hear speech clearly can benefit by combining sound cues from the implant with lipreading and other cues to improve their ability to communicate. In many cases, however, these enhancements require brief or sometimes extended training for the benefits to be realized.

Although cochlear implantation is the state of the art in neural prostheses, it has problems that suggest some general issues in neural implantation. The surgery can produce undesirable side effects: dizziness, because the inner ear is also the organ of bodily balance; infection at the incision site; and occasionally, facial paralysis. Furthermore, the results don't come anywhere near the quality of natural hearing. Another, subtler, potentially troubling problem for implants in general is a hint of an isolating effect that foretells what truly extensive bodily modifications might entail. Some implantees call the quality of the sound they hear "artificial" or "robotic" and, in a surprising twist, others report that instead of feeling that they have rejoined a world from which they have been cut off, they feel alienated from both the deaf and the hearing communities, with the implants leaving them in limbo without full membership in either world.

Despite these problems, the general success of cochlear implantation suggests how digital implants might correct other human deficiencies, and even extend normal human endowments. If a cochlear implant can turn physical sound into the sensation of sound in a deaf person's brain, could a retinal implant turn physical light into the sensation of light in a blind person's brain? Even more interesting, if the implant were sensitive to wavelengths of light that humans ordinarily do not see, such as infrared radiation, could it give a person hypervision?

Similar intriguing questions could be asked about "smart" prosthetic limbs, in which sensors would encode information about a limb's position in space and the textures it encounters. The information from the sensors would be changed into neural signals and sent to the appropriate part of the brain, which would respond by providing motor signals to the hand or leg to produce fine movement control. Suppose also that motors and power sources are built into the limb, or even that the neural control is extended to a device outside the body such as an exoskeleton or vehicle. The result would be a person with enhanced strength, speed, mobility, or reach.

Along similar lines are what might be called internal prosthetics; that is, replacements for organs such as the heart and the liver. Artificial hearts have received the greatest attention and have steadily improved, with reduced risks from the implantation process, longer lifetimes once implanted, and a better ability to restore a recipient to something like normal life. Other internal body parts are also under development, and in some cases commercially available, from skin and blood to tiny implanted devices that automatically release insulin for diabetics. If we reach the point where the artificial versions are superior in capacity or lifetime to natural organs, we might realize the dream of extending the human lifespan by bionic means.

As we imagine the Six Million Dollar Human coming into being through these physical prosthetics, we can also imagine mental prosthetics that go beyond merely injecting electronic pulses into the brain. Such enhancements might, for instance, give the brain additional capacity by holding data in an exterior module, retrieving it on command, and recording whatever experiences are worthy of permanent storage. Or they could give the human brain new levels of computing power, or enable direct brain-to-brain or brain-to-machine communication. Another approach might be to use chemical rather than neuroelectronic means to alter brain function. At least one company is developing an implantable chip that contains several hundred minute reservoirs that can be filled with any desired set of drugs, to be dispensed to the body in variable combinations and dosages under microprocessor control. Although the immediate medical purpose of the device is to deliver therapeutic drugs, there is obvious further potential to modify mental acuity, mood, and personality.

These bionic possibilities require technological advances at every level, as I will discuss later in this book, because the obstacles are formidable. For example, despite the improvement offered by cochlear implants, fully replicating human hearing is an enormous task; the current technology activates only a small fraction of the sensors in the inner ear, the 15,000 hair cells in the cochlea. Consider then what it would take to achieve a reasonable artificial version of human vision, which employs 130 million rod and cone sensors in each eye. There are problems with physical implants as well, and not only the difficult issue of linking a synthetic leg to a brain. If they are to break Olympic running records, runners equipped with bionic legs will need power sources that are more long-lasting, powerful, and compact than present-day batteries.

FEELING BIONIC

But to motivated physicians, engineers, and scientists, these barriers are there to be broken, and to them and humanity in general, any technology that eases suffering by repairing or replacing physical damage should be pursued. Nevertheless, there are legitimate questions, including moral issues, about the wisdom and desirability of bionically modifying people. On the purely medical side, the unwanted possibilities include some already noted, such as infections from the implant process and other harmful effects that might develop over time.

Even if we can avoid undesirable physical effects, bionic modifications might have unwanted psychological outcomes or, expressed more poetically, implantation might damage the human spirit. These problematic effects could include a sense of alienation, such as reported by some cochlear implantees, but the jury is still out on this issue because other implantees have not suffered such strong reactions. For example, the journalist David Beresford, whose severe symptoms of Parkinson's disease have been largely relieved by a neural implant, recently wrote,

> And then there is the psychological side: what is it like to be a 21st century cyborg, with wires coming out of my skull? When I think of it—which is not often—the thought of a wire running deep into my brain is vaguely unsettling, nothing more.

Alternatively, unwanted psychological changes might arise from implants that directly impact the brain in the form of neuroelectronic connections or drug-delivery systems that alter emotional states. To the implantee, such reactions would appear as subjective feelings whose effects would be difficult to evaluate by external diagnosis—another complication when weighing the benefits and drawbacks of changing people in this way.

The potential side effects of implantation require long-term study, only now becoming possible, for example, with a new population of cochlear implantees. Because severe hearing loss can be diagnosed at an early age, cochlear implants have been placed into children as young as 9 to 12 months. These implantees are the first generation to grow up with neurobionic additions, giving researchers the opportunity to better understand the long-term effects of implants and what it means to be bionic. As an example of the problems that might arise, in the summer of 2003, the FDA and the Centers for Disease Control and Prevention presented research showing that children with certain types of cochlear implants are at increased risk of developing a particular form of meningitis. (However, no one knows yet if the implants are responsible, or if children who are good candidates for implantation also happen to be naturally susceptible to the disease.)

Even if bionic additions and implants are proven to be medically and psychologically safe, other questions remain, such as who should have access to the benefits of bionic alteration? If implant technology can extend life, or enhance mental or physical capabilities, how do we decide who receives these precious gifts, and who does not?

The idea that ethical issues might surround bionic modifications, especially cognitive ones, that one might think are purely beneficial, might seem far-fetched. We are not yet, and we might never be, able to modify people sufficiently to change their mental nature. Some researchers believe that artificially enhanced natural minds, and fully synthetic ones, will prove impossible to achieve. This question engages philosophers, psychologists, and cognitive scientists as well as robotics experts: Can we really build artificial brains and link them to artificial bodies? And even harder to answer is the question: If an artificial brain can be built, is the result a self-aware mind, like the one with which we humans experience our own consciousness? The next chapter addresses this complex and perhaps unanswerable question, when we begin to consider artificial beings as they exist today.

Part II

How Far Along Are We?

Can machines live? The answer is yes in the virtual history of artificial beings, but we don't yet know the answer in reality. We have progressed enormously in building artificial bodies, sensory apparatus, and brains, and the progress is accelerating. To understand where we now stand requires insight into today's technology, but first, it requires consideration of an issue beyond engineering: What do we mean by the brain and the mind, and how do they connect to the body?

5

Mind–Body Problems

In classical philosophy, there is only one Mind–Body Problem, that is important enough to be capitalized, but in the world of artificial beings there are several. The philosophical version is an old metaphysical issue, easy to state and hard to resolve: What is the nature of the mind, and how does its apparent insubstantiality relate to the materiality of the body? We know they are connected, because each of us continually experiences their interaction within our own private consciousness. Formulate in your mind the intention to pick up a glass of wine, and your hand carries out the action even as you think it; kick a wall in frustration, and your mind registers the sensation of pain. But how does the immaterial mind cause your hand to move as you desire? Why does it turn a neural signal from your foot into the feeling "it hurts"? Indeed, what is it in you that wishes to drink that wine, and directs your body to act accordingly?

For a long time this problem of consciousness was the province only of philosophers. Because of its internal, subjective nature, consciousness has seemed a difficult subject for scientific study, although some efforts were made in the nineteenth century. Writing in 1890, William James, a founder of modern psychology, concluded that consciousness is a process requiring both memory and the selective placing of attention. But for decades after, psychology was dominated by the objective study of behavior—that is, measuring responses to stimuli—rather than the examination of inner states, and insights like James's were not translated into programs of scientific research. Now, however, with new techniques to simulate the brain, and examine it as it thinks, we might be able to understand consciousness on a scientific basis.

The makers of artificial beings are typically neither cognitive scientists nor philosophers, but aspects of the old mind–body query appear in what they do. One issue of immediate practical importance is the actual link between mind and body. Not that the linkage is always necessary: A body alone, or a brain alone, is enough for some purposes. The designers of programmed animatronic entertainment robots need only bodies that can be fully controlled without any built-in intelligence; researchers in AI would be delighted to produce a brain that shows a high level of intelligence without bodily attachments. But a fully functional artificial creature needs both brain and body, connected so that the brain controls the body and the body informs the brain, and bionic humans need linkages between their brains and artificial limbs or other devices.

THERE ARE NO EASY ANSWERS

Connecting the mental to the physical adds a layer of complexity. The engineering of such connections is the first mind–body problem for artificial beings, and in a way, the least troubling—not that the solutions are easy, but at least there can be agreement about the need to design appropriate interfaces between artificial brain and artificial body, or between a human brain and a mechatronic system (as defined earlier, a device that combines mechanical and electronic elements) such as a prosthetic limb. The human–mechatronic interfaces are the more difficult and involve medical considerations as well, but although there are practical and ethical issues, they do not seem to represent deep philosophical divides.

There are, however, profound differences of opinion on two other questions about artificial creatures that are linked to the mind–body quandary. They generate considerable controversy and the answers might determine the eventual success of the entire enterprise of building artificial creatures. The questions are:

- Can an artificial brain support a conscious artificial mind, as the human brain does human consciousness?
- Is it necessary to embed an artificial brain in a body for the brain to become fully intelligent, functional, and perhaps conscious? As a corollary, might a synthetic body be enough to imbue an artificial mind with a high order of intelligence?

Both questions arise because in the recipe for an artificial being, which reads "one part physical, well mixed with one part mental," we know little about the second ingredient compared to the first, and hardly know how to stir the ingredients together, because we do not know our own recipe—though we've sought it for a long time. One formula goes back to René Descartes in the seventeenth century. He made consciousness central when he stated, "I think, therefore I am," and went on to reason that humans have a dual nature. People, he wrote, are like animals in that both are flesh machines built of matter, which is defined by its extension in three dimensions: but humans have an additional facet, mind, defined as the ability to think. What Descartes could not explain to anyone's full satisfaction, however, was how the two categories interrelate, although he attempted to localize that interaction in the pineal gland.

The dualistic idea that human existence includes an intangible part still carries power in religious and spiritual traditions that hold that an immaterial soul survives the death of the body. And it carries enormous weight for each person. Each of us, looking within, feels that *something* is going on internally that has a different character than the physical operations of the body—call it soul, personal identity, or what you will, it is the core from which each of us gazes out into reality.

However, most contemporary cognitive and neural scientists would say that the mind is the result of physical processes in the brain and hence has a material basis. The Nobel Laureate Francis Crick, who codiscovered the structure of DNA with James Watson and Maurice Wilkins, represents this view. His 1994 book *The Astonishing Hypothesis: The Scientific Search for the Soul* opens with,

> The Astonishing Hypothesis is that "You," your joys and your sorrows, your memories and your ambitions, your sense of personal identity and free will, are in fact no more than the behavior of a vast assembly of nerve cells and their associated molecules.

While scientists accept that the mind arises from the material operations of the brain, this does not solve the classic Mind–Body Problem but it does change its formulation. In modern terms, the question becomes, How can we understand consciousness in scientific terms? Or to put it more specifically, What is the exact nature of the link between physical and chemical activities in the brain and each person's internal sense of consciousness?

This question has several answers of varying degrees of difficulty, as noted by David Chalmers, a philosopher at the University of Arizona. Some aspects of consciousness, such as the ability to choose among and react to external stimuli, are unquestionably susceptible to scientific explanation, though it will take years of effort to understand them. But the aspect that Chalmers calls the "really hard problem" is this: Why do we have a varied internal life at all? Every function of consciousness that supports the physical operations of the body would serve us equally well without these subjective experiences, and so, as Chalmers says, "it seems objectively unreasonable" that we should have them, and yet we do. No one knows why, and this is why people speak of the "mystery" of consciousness.

Although these are profound questions about our own nature, they are closely linked to AI and artificial beings because modern cognitive science is partly inspired by computational science. The exploration of machine thinking has provided significant and useful metaphors for human thought since the 1960s—not long after Alan Turing's seminal 1950 paper—when psychologists and cognitive scientists began using computers to model human mental processes. Conversely, those who want to build machines that think are inspired by the science of natural thought, so the problem of consciousness is deeply important for both groups.

The basic issue is that although we know a great deal about the brain, we know far less about its intangible correlate, the mind. The brain, after all, is a working physical part of the body, like the liver or heart, whose physiology and functions can be studied. In a typical adult, it is a 1.3 kg (3 lb) mass of tissue that contains about 100 billion neurons and supporting structures. Like any other part of the body, it uses energy and requires nutrients. Through dissection and other techniques, we know its complex anatomy, from the cerebrum with its two walnut-like halves, to the brain stem that exits through the lower skull to become the spinal cord. We know the general functions of its parts, and we can identify areas that control bodily movements, process visual information, deal with language, and so on. We know the structure of neurons, and something about how they communicate among themselves and their interconnections in the brain, which can change as a result of experience.

Certainly, further insight is needed. That should include, for instance, fuller knowledge of neurotransmitters—the chemicals like serotonin that carry signals among the brain's neurons by electrochemical means—and more extensive mapping of the brain's functions, especially those like memory that seem to integrate information from different areas. However, scientists firmly believe that their understanding of the brain will steadily grow through the use of electroencephalography (EEG) and the study of the effects of brain damage, and especially through the new techniques of functional magnetic resonance imaging (fMRI) and positron emission tomography (PET) scanning. Both make it possible to observe something never seen before—the operations of a living, working brain.

But that's the brain. The human mind, or human consciousness, is different. We know it is correlated with the brain, because if you cut off certain brain functions, consciousness flees, but we do not understand its nature and origins, largely because it is a subjective experience that is difficult to explore by the objective means that illuminate the brain. (In animals less developed than humans, "consciousness" might be limited to the ability to sense stimuli and respond directly to them. My use of the term goes beyond that baseline level to include human thinking, feeling, and self-awareness.)

The subjectivity of internal experience leads to what philosophers call the "problem of other minds": in principle, we can never truly grasp the nature of anyone else's inner life. In this view, though it is a chilling thought, we cannot be sure that other people have inner lives at all. They might be zombies—behaving like humans, but lacking internal experiences, including emotions and feelings. Regardless of this philosophical point, of course, we all go through life assuming that other people are much like us inside, but the idea of zombies is less far-fetched than we might think. Brain injuries can cause the loss of certain emotional reactions, and psychiatric practice recognizes zombie-like characteristics in some people, who are known as sociopaths. Their actions seem to be appropriate expressions of normal feelings, but they are only playacting, because inside they are devoid of compassion or empathy for others.

The problem of other minds illustrates the difficulty of unraveling consciousness by scientific means. As the neurologist Antonio Damasio puts it: "How can science approach interior phenomena that can be made available only to a single observer and hence are hopelessly subjective?" But now, it seems that brain activity can be made widely observable and linked to interior states through such means as PET and fMRI. We can begin to deeply explore what has been called the last frontier of neuroscience, and the philosopher John Searle, of the University of California, Berkeley, declares to be "the most important problem in the biological sciences"; namely, "How exactly do neurobiological process in the brain cause consciousness?" This question is equally important for the cognitive science of humans, and of artificial beings.

DUELING THEORIES

Despite much intense thinking about how we think, there is still no single theory to explain how the actions of an intricate neural array turn into the deeply felt sense of self we each carry, or that could form a blueprint for an artificial mind. Cognitive theorists, neuroscientists, psychologists, philosophers, experts in AI all have their approaches, showing that the question has yet to be answered to everyone's or even anyone's full satisfaction. What most theories have in common is the attempt to show how neurons work together to give unified perceptions and thought processes, leading to a coherent sense of consciousness. In visual cognition, the mind's need to bring together different aspects of a seen object into an integrated perception is called the "binding problem." Some theories hold that consciousness arises from a greater, more inclusive binding process. Others consider it an emergent property, meaning that although it can be traced to neuronal behavior, no single neuron is conscious, nor can a simple sum of all the neuronal properties account for consciousness.

These theories cover a wide range—from the view that the mere operation of the parts of the brain constitutes consciousness, to the belief that consciousness arises from as-yet-unknown natural phenomena, to the extreme view that the human mind can never fully understand itself. The unsettled nature of the field, and the lack of more than the beginnings of hard data, is shown by the disputes among proponents of different theories, disagreements often relying on assertions that depend on key words like "consciousness," "intentionality," and "meaning." Because these words are hardly rigorously defined, the quarrels often represent no more than differences of opinion or interpretation, producing much waste heat and little useful scientific light. Nevertheless, there are nuggets of truth among these conflicting ideas.

The most startling view is that consciousness is illusory, or at least behaves very differently from our internal sense of it. Many people, whether philosophers and scientists or not, find this approach not only counterintuitive but repellent, because it violates cherished

beliefs about personality and free will. In his 2002 book *The Illusion of Conscious Will,* the psychologist Daniel Wegner gives experimental evidence about the relation between a person's sense of volition—which leads to a bodily action like reaching for a wine glass—and the neural impulse that actually moves the hand. The unexpected result is that the decision to move does not necessarily precede the motion. As Wegner puts it, "It usually seems that we consciously will our voluntary actions, but this is an illusion. . . . Our sense of being a conscious agent who does things comes at a cost of being technically wrong all the time." He goes on to argue that our experience of conscious will nevertheless makes us feel that we are beings who can make moral choices, but his results tend to undermine bedrock assumptions about human choice and responsibility for our actions.

The cognitive theorist Daniel Dennett of Tufts University takes an even stronger view of consciousness as illusion, as articulated in his 1991 book *Consciousness Explained*, and other writings. According to Dennett, what goes on in the brain is distributed cognition, a complex pattern of events occurring at different times and at different physical sites in the neural array. Thought dispersed temporally and spatially is a far cry from Descartes's idea that the center of the self resides in a single location, and eliminates the idea of a physical core for consciousness. Taking the argument further, Dennett believes that there is no central core of any kind for personhood. Self-consciousness, he says,

> is that special inner light, that private way that is with you that nobody else can share, something that is forever outside the bounds of computer science. . . . That belief, that very gripping, powerful intuition, is in the end, I think, simply an illusion of common sense . . . as gripping as the commonsense illusion that the earth stands still and the sun goes around the earth.

Instead, he says, "you *can* imagine how all that complicated slew of activity in the brain amounts to conscious experience . . . the way to imagine this is to think of the brain as a computer of sorts." (Italics in the original.)

If Dennett downplays the strong internal sense of our own consciousness, the philosopher John Searle gives great weight to those

same internal feelings. In his 1997 book *The Mystery of Consciousness*, Searle takes the experience of consciousness as a core reality precisely because it is an unmistakable interior event. His rebuttal of Dennett's ideas is curiously reminiscent of Descartes's "I think, therefore I am." Searle writes,

> But where the existence of conscious states is concerned, you can't make the distinction between appearance and reality, *because the existence of the appearance is the reality in question.* If it consciously seems to me that I am conscious, I am conscious . . . it is just a plain fact about me—and every other normal human being. . . . (Italics in original.)

Searle does not use this perspective to build a theory of consciousness, but Francis Crick explores such a theory in detail. In *The Astonishing Hypothesis* and elsewhere, and with his colleague Christoff Koch, he approaches the phenomenon through the binding problem in visual cognition. This particular function of mind draws on a large fraction of the brain, where certain groups of neurons deal with specific parts of what we see, such as color, movement, and the edges of objects. The mind brings these elements together to produce an integrated visual understanding that is an important part of our mode of thought. Using a variety of evidence, Crick concludes that binding of this sort is produced by neurons located in different and specific parts of the brain that fire in a synchronized way, on average 40 times a second. He does not claim that this conclusion solves the problem of consciousness, but believes that the full answer must begin with just this kind of consideration of enormous numbers of neurons operating together.

The Nobel Laureate neuroscientist Gerald Edelman, of Rockefeller University in New York City, and his colleague Giulio Tononi also consider the unified action of groups of neurons, most recently in their 2000 book *A Universe of Consciousness: How Matter Becomes Imagination.* Their theory draws on evolutionary development, which, they say, has formed our brains to process information more powerfully than human-made computers can. A kind of Darwinian survival of the fittest affects individual brains as well, through neuronal group selection: As a brain develops, the groups of neurons that survive are those that respond well to stimuli. They represent conceptual categories, and through the process of "reentry," constantly trade information back and forth as if the brain were talking to itself.

Edelman and Tononi conclude that interactions between two particular structures in the brain are mostly responsible for consciousness: the cortex or gray matter—the outer layer of neurons that deals with sensory impulses and higher mental functions—and the thalamus—a part of the brain associated with emotion. Moreover, there are two levels of consciousness. Primary consciousness is perceptual awareness of the world in the present, but it is not consciousness of self. That level comes with higher-order consciousness, which depends on language and on social interactions and which has knowledge about the past and future as well as the present; it is what humans add to their primary consciousness.

The physician and historian of ideas Israel Rosenfeld also believes in the importance of coherence over time, the sense of self we maintain as a continuous internal presence throughout our lives, or at least our adult lives. As William James saw more than a century ago, this long-term coherence is a function of memory, and Rosenfield emphasizes that "consciousness and memory are in a certain sense inseparable, and understanding one requires understanding the other." But how does this continuous memory develop? According to Rosenfield, memory is created because the brain resides in a body:

> My memory emerges from the relation between my body . . . and my brain's "image" of my body (an unconscious activity in which the brain creates a constantly changing generalized idea of the body . . .). It is this relation that creates a sense of self.

None of these approaches is a definitive explanation of consciousness that is supported by complete scientific evidence. It can be argued also that none truly confronts the hard problem of subjective experience and why we have it, at least not within the framework of what we know about the brain. Edelman and Tononi touch on this issue when they write,

> while we can construct a sensible scientific theory of consciousness . . . that theory cannot replace experience: Being is not describing. A scientific explanation can have predictive and explanatory power, but it cannot di-

rectly convey the phenomenal experience that depends on having an individual brain and body.

Some thinkers feel that explanations are beside the point anyway, believing that our mental functions—such as using categories to make sense of the world—are innate and cannot be approached by the tools of cognitive science. At least one thinker, however, believes that an explanation is possible, but only by drawing on new phenomena. That tack is taken by the Oxford University mathematical physicist Roger Penrose, as expressed in the subtitle of his 1996 book *Shadows of the Mind: A Search for the Missing Science of Consciousness*, and in his earlier writings.

Penrose does not deal much with neurons and neurobiology. He begins with a famous mathematical proof called Gödel's theorem. This result, derived by the Austrian-born mathematician and logician Kurt Gödel in 1931, is of prime importance in modern mathematics. It proves that any formal system—such as the set of axioms that defines classical geometry, or a computer program—can logically generate statements that are true, but that cannot be proven within the system. Gödel's proof implies that there are true mathematical results that cannot be derived by computers, which operate by strict logical rules, but can be derived by humans.

Thus, concludes Penrose, the human mind supplies something extra, something "noncomputable" that lies beyond what computers can do. This quality, Penrose asserts, arises from phenomena at the microscopic quantum level, where everyday laws of cause and effect are replaced by laws of probability. He suggests that a new kind of quantum behavior in the brain, perhaps "quantum gravity," provides this essential element of noncomputability—although the details of this novel quantum physics are as yet unknown. But neurons are too big to follow the quantum laws, and so Penrose speculates that consciousness arises in smaller structures in the brain called microtubules. Because Penrose hypothesizes that consciousness comes from new natural phenomena without any evidence that these exist, his ideas have been much criticized.

PEOPLE THINK, BUT DO DIGITAL CREATURES?

Apart from the merits or deficiencies of Penrose's approach, it illustrates one of two main corollaries that accompany theories of consciousness; namely, that machines can never think or be conscious in the way that people are—accompanied, of course, by the conflicting belief that machine consciousness is possible. In the early days of AI, the answer seemed simple. The pioneering AI researchers considered thinking to be the processing of information, which is, in turn, the manipulation of symbols; hence, minds are simply systems for processing symbols. As it happens, our own minds are based on brains made of neurons, but the physical nature of the processor is unimportant. Thus, whether the "brain" consists of billions of living nerve cells, a stack of silicon chips, or for that matter, one of Isaac Asimov's positronic units, the important thing is that symbols are meaningfully manipulated. When that happens, thinking, and perhaps even consciousness, occurs.

This view is often called, semijocularly, GOFAI—"good old-fashioned artificial intelligence"—and is now recognized as falling short of a complete approach to machine intelligence. Decades ago, as computer programs began to manipulate symbols in meaningful ways such as carrying out mathematical proofs, proponents of GOFAI felt we were well on our way toward full AI. But as understanding grew, we came to realize that GOFAI omits some aspects of cognition—for instance, the sensory experience of smell—which might not be represented by words or other symbols inside our minds.

Today, with AI and cognitive science far more advanced, and theories of consciousness abounding, there is ammunition for those who believe that machines can think and for those who don't. In Daniel Dennett's view, a human mind that is thinking is running what amounts to a computer program that processes information. To Dennett, this scenario opens the door to machine thought. He claims that "a computer that actually passed the Turing test would be a thinker in every theoretically interesting sense," and adds "I do think it's possible to program self-consciousness into a computer." But

Roger Penrose would insist that computers can never do all that human minds can, nor even simulate those activities. John Searle also believes that it takes more than mechanical computation to constitute thinking. He calls the belief that computation is the same as thinking "Strong AI," and rejects this in favor of "Weak AI"—while computers can simulate human thought, the simulation of thinking is not necessarily thinking.

In 1980, Searle gave what is probably still the best-known rebuttal to Strong AI, the "Chinese Room" scenario, which emulates how a computer works. Imagine that you are asked to answer questions presented to you in Chinese, although you speak only English. You sequester yourself in a room containing many tiles marked with Chinese symbols (the database) and a book of rules written in English (the program). Questions, written in Chinese, are presented through a small slot (input). You (the CPU, central processing unit) match the incoming Chinese characters to entries in the book and then manipulate the Chinese character tiles as the book directs. That leads to new Chinese characters, the correct answers to the questions, which you present to the world through another slot (output).

The heart of Searle's contention is that although this process enables you to obtain correct answers, in no way do you understand Chinese as you obtain those answers. The distinction is between what a computer does, which is to manipulate formal symbols like Chinese characters, and what our minds do, which is to add meaning to the symbols. Hence, concludes Searle, although his hypothetical computer passes the classic Turing test administered in Chinese, "programs are not minds," and a computer or robot can never be conscious.

Many serious objections have been raised to the Chinese Room. One counterargument holds that, whether the person in the room understands Chinese or not, the system as a whole—database, CPU, and so on—does. Other reactions pit the scientific stance against the philosophical one, a common theme in consciousness studies (in my opinion, the answers will come from science, but the philosophical questions are invaluable in pinpointing the issues). Dennett, for instance, warns that the Chinese Room acts to dissuade people from imagining in detail how proper software design could engender machine consciousness, in that it makes a flawed analogy that manipulates our intuitions. Disputes like these indicate that expectations for machine consciousness still rely more on opinion than on fact. But why should this be surprising? Apart from some conscious functions we share with animals, our best and sole model for an artificial mind operating at a high level is the human mind itself—and we do not know that very well.

To be realistic, the question "Can machines think?" is of limited pragmatic interest at this point. We are only in the earliest stages of creating machine intelligence; meanwhile, useful creatures are being created without their builders taking a stand or caring whether they are conscious or "really" think. Looking to the future, however, many researchers believe that machine consciousness will be realized. Some, such as the roboticist Hans Moravec of Carnegie Mellon University, adopt the visionary view that artificial minds will surpass ours. In his 1999 book *Robot: Mere Machine to Transcendent Mind,* Moravec predicts that "Fourth-Generation Universal Robots" will be available around the year 2040, with "human perceptual and motor abilities and superior reasoning powers," and suggests that we humans are "parents [who] can gracefully retire as our mind children grow beyond our imagining."

Whether or not this particular prediction is correct, it is true that as artificial brains and creatures become more capable and enter human society, the question of their consciousness becomes more pressing. The practical reason to be concerned is that unless the being is truly conscious rather than only seeming so, it might make faulty decisions—perhaps dangerous ones—in dealing with humans. To know that misapplication of its strength could harm a human, the artificial being might need to develop empathy, through the sharing of such human feelings as the sensation of pain; otherwise, it might become a high-tech sociopath.

From the human perspective, there is a moral issue as well, because once an entity crosses a certain threshold of sentience, we enter into a different relationship with it. No one hesitates to kick a rock, but some of us balk at uprooting a plant; most people who gladly swat a fly would never hurt a cat or dog. Similarly, we would feel differently toward a machine without a shred of consciousness than toward an artificial being we know to have inner feelings.

And finally, there is a reason to pursue the possibility of artificial minds that carries broad scientific value: By contemplating what artificial consciousness means, and from attempts—however ill-defined and halting—to build creatures with minds, we learn about our own minds. In the eighteenth century, Jacques de Vaucanson hoped to build a synthetic human body so detailed that it would teach us about our own bodies. Now we have a similar possibility for our minds.

I AM, THEREFORE I THINK

In contemplating the possibility of an artificial being with an artificial mind, we must recognize that the mind is contained in a real, physical body. Many ideas and debates about machine thinking assume that it arises as a disembodied intelligence within a computer. Artificial creatures, however, are different. They need to think, yes, but that ability must be coupled to interaction with the world: sensing it in various tangible forms rather than symbolically, assessing that flow of data, and deciding how to respond with physical action. The decision can be direct and immediate, though not necessarily simple, as in a robot choosing where to put its feet so as to walk in a given direction. At higher levels, the sensory input, processing, and decision making might reach the sophistication of navigating through a complex environment, or conversing with a person in a human way—that is, passing the Turing test, not as a presence hidden behind a screen, but by actually being there, to listen and speak.

In short, artificial beings are *embodied* intelligences. To some researchers, that mind–body association is the key to making fully successful creatures. The difference between this approach and approaches based on disembodied intelligence remains controversial. It is why Rodney Brooks's construction of Genghis, his legged robot that learned to walk by using a distributed, reactive intelligence rather than a central symbol-oriented intellect, was revolutionary. Genghis's success challenged approaches such as that used for Shakey, the prototype for a proposed battlefield unit, which proved an unworkable example of GOFAI. Now the idea of embodiment is at the core of one approach to the design of intelligent mobile beings.

Rodney Brooks's experience with Genghis, Cog, and other robots has made him a leading proponent of the significant interaction between synthetic body and artificial mind. His beings are built with two central principles in mind. One is situatedness, meaning (as Brooks defines it),

> the creature or robot is . . . embedded in the world . . . [it] does not deal with abstract descriptions, but through its sensors with the here and now of the world, which directly influences the behavior of the creature.

The other is embodiment, meaning that,

> the creature or robot . . . has a physical body and experiences the world, at least in part, directly through the influence of the world on that body.

As examples, Brooks points out that a computerized airline reservation system is situated but not embodied: It deals with the outside world, but only by means of messages. An assembly-line robot that spray-paints cars, however, is embodied but not situated: It has a physical presence that accomplishes a real task, but makes no judgments about the cars it paints, and is unaffected by them, simply repeating the same actions over and over.

Brooks foresees a situated robot with a well-equipped body that could develop a conceptual understanding of the world in the same way we do. In 1994, he proposed that a humanoid robot with capabilities including vision, hearing, and speech, and the ability to physically manipulate objects, would "build on its bodily experience to accomplish progressively more abstract tasks." This possibility is supported by ideas from cognitive science, such as Israel Rosenfeld's approach, which gives great weight to the physical body in determining memory and consciousness.

The cognitive scientists George Lakoff and Mark Johnson are even more specific. In their 1999 book *Philosophy in the Flesh: The Embodied Mind and its Challenge to Western Philosophy*, they postulate that the high-level functions of mind, such as language, begin as metaphors for how our bodies interact with the world. "The mind is inherently embodied," they write, adding, "Thought is mostly unconscious. Abstract concepts are largely metaphorical." Reason itself, they believe, is intimately connected with our physical nature:

> Reason . . . arises from the nature of our brains, bodies, and bodily experience . . . the very structure of reason itself comes from the details of our embodiment. The same neural and cognitive mechanisms that allow us to perceive and move around also create our conceptual systems. . . .

But although Genghis learned to walk, and Brooks's robot, Cog, seemed alive when it turned toward a visitor, embodied intelligences have yet to demonstrate that they have developed higher functions of mind operating at abstract levels. Numerous questions remain about this approach. The pioneering AI researcher Marvin Minsky, for instance, has called emphasis on robots "unproductive" and "bad taste on the part of my fellow scientists," adding,

> in the 1950's and '60's . . . we found, OK, you can build a robot and have it walk around the room and bump into things and learn not to, but we never got any profound knowledge out of those experiments.

Despite such sharp differences of opinion, researchers continue to attack the mind–body problems for artificial beings on many fronts. Some research efforts focus on the pragmatic goal of developing operational creatures; others operate on a deeper level that hopes to build fully conscious beings. Mind–body considerations apply also to bionic humans or cyborgs; for instance, the different subjective reactions that have been reported by the recipients of cochlear and brain implants, some of whom are troubled by a sense of isolation or strangeness and some who are not. There is evidence as well that neural implants cause actual changes in the brain and the way in which it perceives the body. This is a function of the brain's plasticity, the change in its neural arrangement as a result of external influences. The changes caused by a neural implant that controls an artificial limb or external device are likely to be beneficial toward incorporating that nonliving addition into a person's body image. Still, if altering people from fully natural to partly artificial literally changes their minds, that presents another mind–body problem, one with potentially serious ramifications.

The hard problems of consciousness remain hard. The debates over mind, thought, and consciousness might continue for a long time or might never be resolved, either for ourselves or for artificial beings. For our own constructed creatures, suggests Rodney Brooks, the only answer we might be able to trust is the one we trust for ourselves:

> Perhaps we will be surprised one day when one of our robots earnestly informs us that it is conscious, and just like I take your word for your being conscious, we will have to accept its word for it. There will be no other option.

However, although the full mind–body recipe remains unknown for us and our artificial kin, a great deal of progress has been made on the bodily ingredient, as the next chapter shows.

6

Limbs, Movement, and Expression

Those early Greek theatrical simulations of living beings conveyed a sense of life through motion, and motion remains a hallmark of artificial beings. We feel that motion means vitality, and hence is essential for lifelike synthetic creatures; in fact, the Nursebot, a robot developed at Carnegie Mellon University, which is designed to assist the elderly, not only has facial features (cartoonish ones) but blinks its eyes at regular intervals so that observers understand that something is "alive" in there, even when the robot is at rest. There are also practical reasons to incorporate mobility: What would be the use of a household robot that could not clean the floor, or an industrial robot that could not move its arms to weld an automobile door panel? While an artificial mind is an essential part of a useful robot or android, it is the addition of motion along with sensory interaction that turns a mind into a full being.

But what kind of motion? That depends on the goal: Is it to make a robot that is functional, but might not look at all human, such as Roomba, a robotic vacuum cleaner? Or is it to make a robot that, although it could not be mistaken for human, is sufficiently humanoid to operate in everyday environments, from homes to offices and factories—the approach taken by the Honda Corporation. The goal might be a true android that, among other humanlike behaviors, walks on two legs and grasps and holds objects, and does so with human-appearing body parts, as might be important for entertainment robots. Or the aim might be to make a being that surpasses human capabilities with enhanced strength or speed, a feature that interests the military, or is designed from an utterly different premise than matching the human body, such as changing its shape to suit the task at hand.

For designers of artificial beings, there is an obvious appeal in selecting the simplest appearance and physical behavior that will do the job. For instance, there are easier means of locomotion than walking. As a kind of controlled fall endlessly caught and repeated, walking requires the ability to sense and maintain bodily balance, which requires in turn appropriate sensors and cognitive ability. But no matter what the design or form of its limbs, specific types of sensing and cognition are necessary if the being is to operate in the real world.

For instance, it would be valuable if the Nursebot robot, assisting in caring for the elderly, could follow a human's instructions such as, "Go to Mr. Smith's room, ask him if he would like lunch, and if he says yes, guide him to the dining room." To successfully navigate its way to a specified location, whether externally determined or self-selected, the being needs vision or other means to examine and map its surroundings, knowledge of its present location in that environment, and the ability to plan a workable route from here to there while recognizing and avoiding obstacles, making proper use of doorways and so on, along the way.

Similar considerations apply to artificial arms, hands, and fingers. They can be correctly positioned to perform their functions only if the artificial being has what is called in humans the kinesthetic sense; that is, the ability to determine where the body part is in space relative to the body itself. And to ensure that a hand is set correctly to safely hold a fragile object, or that an arm exerts enough power to lift a heavy load, the being needs a sense of touch and the ability to sense how much force is required to hold or move an object.

Hence two things are required to give artificial beings the ability to move and to handle objects. One is mechatronic engineering design, which couples the mechanical principles of wheels, gears, and joints to electrical and electronic devices such as computer chips and servos. A servo is a type of actuator that animates artificial limbs. It is similar to an electric motor, with a shaft that can be rotated to any desired angle as set by an electrical signal. The units are powerful for their small size, and because they can be positioned just as desired, are good choices for versatile control of artificial body parts. The characteristic whine that always seems to accompany robotic movement in the movies is the sound of a servo being driven to a specific position.

Equally important, and generally more difficult to construct, are the bodily senses and cognitive abilities that enable the being to work out, for example, where and how its legs must be set to walk in a given direction, and then issue the necessary commands to its servos. That's a complicated task. To make it even more demanding, the computation and the mechatronic response must occur quickly enough to allow the being to function in real time.

Depending on the bodily details, different levels of cognitive abilities are needed. As we will see below, wheeled robots, for instance, and robots that walk on four or six legs need less brainpower than those that walk on two legs.

WHEELED, TREADED, AND TRACKED

Like a tricycle resting firmly on its three wheels, a wheeled being rolling on more than two wheels is inherently stable, and needs minimal cognitive capability to maintain its balance. Wheels offer a relatively simple mode of locomotion that is a good choice for robots when humanoid appearance is neither necessary nor desired. Nevertheless, wheels are not right in all circumstances. They work well on smooth surfaces, but rough terrain or a profusion of obstacles—such as occur in nature, at disaster sites, or in warfare—can defeat them. Under those conditions, continuous treads or tracks with cleats, like those on a bulldozer or military tank, serve better and still provide stable underpinning.

Because wheeled units are comparatively simple to design, they are used as test-beds by those who develop robots. Students and faculty in Carnegie Mellon University's Robotics Institute, for instance, have gotten used to seeing wheeled cylindrical robots rolling through their building, testing the robots' ability to map the environment and navigate through it.

Also for the sake of simplicity, present-day commercially available robots—other than those that simulate four-legged pets—roll on wheels. At least one, the vacuum-cleaning robot called Roomba, placed on the market in 2002, is a consumer item. It is made by the iRobot Corporation whose co-founder, chairman, and chief technology officer is Rodney Brooks, the MIT roboticist responsible for Genghis and Cog. Roomba is a 6-pound, disc-shaped unit, slightly more than a foot across, that rides on two wheels and an agitator brush. Unlike conventional vacuum cleaners that draw their electrical power from a cord plugged into a wall socket, Roomba operates from rechargeable batteries and hence has complete freedom of motion.

Put Roomba down in the center of a room, and it begins to cover the floor in a pattern of broad spiral sweeps (which give Roomba its name) until the robot encounters a wall. Then its behavior changes to track along the wall and clean the wall base with side brushes, until it makes another foray across the room. If Roomba encounters an obstacle such as a chair leg, it stops and goes off in a new direction, and it is also intelligent enough to halt and turn around at the top of a flight of stairs. Eventually the robot cleans the entire area, and then it stops, ready to be placed in another room or recharged. While the robot has sensors and a level of intelligence called "heuristic learning logic" that was first developed for units used to clear minefields, the fact that it does not need to maintain balance as it rolls simplifies its design and lowers its price.

The iRobot company has also developed tracked robots in the form of small units called PackBots. These were designed to function in difficult environments, and so were well suited to enter the wreckage of the World Trade Center after the terrorist attack of September 11, 2001. Equipped with video cameras, they sent back images as they clambered over debris, and helped human searchers recover a number of bodies.

In 2001, in the first actual appearance of robots in warfare, the U.S. military used PackBots in the Afghanistan campaign to search caves for Al-Queda operatives and to locate hidden weapons and mines. These 42-pound, backpack-size units were not truly autonomous because they required human operators; but if they and other military robots are developed further as envisioned, we might one day see a single soldier monitoring several self-directing units, thus achieving the military dream of force multiplication: "You could have 10 people on the battlefield doing what once took 40 soldiers," notes Ronald Arkin of Georgia Tech, who has written software for military robots.

EIGHT LEGS TO TWO

Treads are good for tackling steep grades, but less so for stairs. In fact, for typical human environments that can include stairs or varied surfaces such as a deep-pile rug or a tile floor, neither treads nor wheels work as well as legs and feet, although not necessarily only two of each. Robots that walk on four or more legs remain balanced even if not all the legs are on the ground. That inherent stability can be valuable for the broken terrain to be found in the field, or in NASA-sponsored expeditions of planetary exploration, and requires fewer cognitive resources to maintain balance.

Rodney Brooks's Genghis robot illustrated that multilegged walking on flat surfaces and over obstacles can be achieved with only limited computational power. My I-Cybie robot dog is not the most advanced robot toy on the market and operates at a low level of artificial intelligence; nevertheless, it manages to shuffle along nicely on four legs (it can even right itself to stand firmly on those legs if you tip it over, and can also work its way around obstacles, walking off in a new direction when it bumps into something). One multilegged robot that has dealt with extremely rough terrain is the eight-legged Dante II. In 1994, this unit semiautonomously walked and rappelled its way down into the crater of an Alaskan volcano, which it explored for the better part of a week. Many six-legged robots have been developed as well.

To develop general approaches to designing robots with legs, however, especially those that walk on two legs as people do, is a complicated business that has occupied researchers for years. One well known establishment, the MIT Leg Laboratory, is devoted to the study of locomotion and the construction of legged robots. Other research on two-legged beings is carried out internationally at universities and corporations, with the leading efforts in Japan at Tokyo University, the Honda Corporation, and elsewhere. These efforts are central to the widespread proliferation of artificial beings. If they are to work alongside people and interact with them, they will have to be two-legged and generally humanoid so that they can operate in regular human environments and use human devices such as screwdrivers and doorknobs. Another advantage of humanoid robots is the versatility of the human frame. Cheetahs run faster, dolphins swim faster, and chimpanzees exert more strength than humans, but the multipurpose human frame is moderately good at all these things. This adaptability defines the approach chosen by many researchers to build a humanoid robot with similarly broad functionality.

That versatility begins with the ability to walk on two legs, as shown, for instance, by the robots developed by Hirochika Inoue, of the Department of Mechano-Informatics at the University of Tokyo. His latest units, called perception–action integrated humanoid robots, are named H6 and H7. They are human-shaped with head, torso, and limbs but lack facial features and are recognizably mechanical. Conditioned as we are to seeing inscrutable movie robots stalk off to carry out their plans, like the faceless Gort in *The Day the Earth Stood Still,* we might expect H6 and H7 to look menacing—but they are neither hulking nor brutish. They are white in color, and not very big: H7 stands 1.4 meters (4' 7") tall, with a mass of 55 kilograms (120 pounds, including 9 pounds of batteries to power it).

It is illuminating to see H6 and H7 in action because their walking style is almost tentative. When either unit walks, it is accompanied by human minders, placed so as to catch the robot and keep it from damaging itself if its balancing abilities fail and it topples. Moreover, the walking pace is slow. Most telling of all, H6 and H7 do not swing their arms while walking, as people do, but hold their arms at their sides, bent at the elbow. This makes the robots seem slightly nerdy, as if they were too tightly buttoned up to stride freely down the street.

But even that cautious, somewhat geeky walk, though a far cry from the menacing lurch with extended arms featured in horror movies, is highly significant. Its achievement has required considerable mechanical and computational development. Successful humanoid robots need mechanical frames that are highly flexible, which is defined in terms of "degrees of freedom." Each degree of freedom means the capability to move a limb or other appendage in a given direction about a joint. H7 can move each leg in any of six directions, corresponding to six degrees of freedom, plus a seventh that comes from an adjustable toe joint in the foot. There are 23 additional degrees of freedom built into the robot's body, and all its joints are moved by electric motors that drive gears.

For H7 to walk in a given direction, its legs must be set correctly and their movements coordinated in space and time. At the same time, the robot's body must be constantly adjusted to maintain its balance as it walks. Moreover, the robot cannot be allowed to self-collide—that is, have one moving part strike another, such as the legs becoming entangled. With the robot's humanoid shape and many degrees of freedom, there is a multitude of possible bodily configurations, each of which must be examined to ensure safe and successful walking. This is an extremely demanding computational task that must be performed in real time as the robot advances through the world. The necessary calculation power is provided by the equivalent of two powerful laptop computers built into the robot, with more power coming from other remotely linked computers. Even this computational armory would not be enough without clever algorithms that minimize computation time.

Similar considerations apply to what are probably the best-known walking robots in the world, which have been undergoing development since 1986 at the Honda Corporation in Japan. The earliest unit in this series could indeed have come from a science fiction film. It consisted of a pair of legs attached to a large squarish box the size of a microwave oven, and resembled the walking battledroids seen in the *Star Wars* movies. If this inhuman-looking robot moved with any facility and speed, it would be a fearsome thing to see bearing down on you. Fortunately, this early model was not a very impressive walker: it took all of 5 seconds to calculate the leg position and foot placement for each step, and it could walk only in a straight line.

Further development produced refinement after refinement, but it took 10 years for Honda to unveil its first humanoid walking robots, called P2 and P3, followed by an improved version called ASIMO (advanced step in innovative mobility), announced in 2000. P3 and ASIMO have appeared in Honda's corporate advertising and are available for public events and expositions. In early 2002, ASIMO rang the opening bell at the New York Stock Exchange to celebrate the twenty-fifth anniversary of Honda's listing on the exchange. In fact, the robot is being groomed as a general-purpose unit: According to Honda, "In the future, we anticipate ASIMO developing capabilities in areas such as household assistance and tasks dangerous for humans—like firefighting."

Both P3 and ASIMO resemble a person in a white spacesuit or suit of armor, topped by a helmet with a dark visor; nothing like a face is visible. Each robot carries a sizable backpack, which houses its on-board computer and its batteries. The differences between P3, and ASIMO, developed only a few years later, illustrate the rapid pace of robotic improvement. P3 is the size of a very small adult, standing 1.6 meters (5' 1") tall, but despite the use of the light metal magnesium in its construction, comes in at a hefty 130 kilograms (285 pounds). ASIMO, however, has been pared down to child-size, standing 1.2 meters (3' 10") tall and weighing only 43 kilograms (95 pounds) Yet this smaller robot is smarter and more able than its older brother, although it walks slightly slower, at 1.6 kilometer per hour (1 mile per hour). One battery charge keeps it going for 30 minutes.

ASIMO is more flexible than P3, sporting 26 degrees of freedom including six in each leg. In addition, it incorporates a "predicted movement control." Like a walking man seeing a corner coming up and shifting his stance to accommodate a turn, ASIMO looks ahead to the next stage of its motion and shifts its center of gravity accordingly. With its greater flexibility and improved cognitive skills, ASIMO is a smoother and better walker than P3.

ASIMO can maintain its balance standing on a steeply tilting seesaw, using a telescoping knee joint rather than bending a knee as a human would. It can neatly execute a turn (which P3 does only awkwardly in a series of shuffling steps), balance on one leg, and climb confidently up and down stairs, although it must know the stair height in advance. In fact, to a human observer, both ASIMO and P3 radiate a certain self-assurance in walking that H6 and H7 do not. The reason is that the Honda robots swing their arms in human fashion while walking, giving the impression of a confident robot that knows what it's doing and where it's going—one more illustration of the effectiveness of humanlike clues built into artificial creatures.

CRAWLING AND MORPHING

Although humanoid robots offer great versatility, and walking on two legs is an important achievement, there are reasons to consider other body shapes and ways to move. Many living creatures progress by crawling, slithering, or skittering rather than walking. For all its versatility, a humanoid body cannot efficiently emulate these motions, which might prove to be the best choice for certain applications. Hence some roboticists are developing robots with nonhumanoid body shapes and means of locomotion.

The robots made by Shigeo Hirose of the Tokyo Institute of Technology, for instance, are not humanoid. His ACM R-1 (active cord mechanism, revised model) is long, skinny, and snakelike and slowly slithers along at about 40 centimeters a second (just less than one mile per hour). Hirose was, in fact, inspired by studying the movements of snakes, and his choice of body type also fits into an engineering philosophy that takes the simplest solution to be best. Instead of trying to make a multipurpose humanoid robot, this approach chooses from among the multitude of possible bodily designs the best one to do a specific job. ACM R-1, for instance, is ideally shaped to explore underground pipes, if not for much else.

While Hirose's robots are not humanoid, they maintain a fixed form. But there is a more radical approach to locomotion and bodily design: robots with no permanent legs or arms, no fixed bodily configuration, that dynamically change their shape and means of locomotion to meet the needs of the moment. One version of these reconfigurable robots, called PolyBots, is being developed by Mark Yim, Ying Zhang, and David Duff of the Palo Alto Research Center. These researchers contrast the fixed assembly-line world of industrial robots with the shifting demands and terrain of the real world and envision a robot that could

> shape itself into a loop and move by rolling like a self-propelled tank tread; then . . . form a serpentine configuration and slither under or over obstacles . . . then . . . "morph" into a multilegged spider, able to stride over rocks and bumpy terrain.

The key to this flexibility is to construct the robot from individual modules of only one or two types, but numbering in the hundreds and potentially in the millions.

One PolyBot under study, dubbed G2, has modules, each a cube 5 centimeters (2 inches) on a side, that can automatically connect with each other to form long strings. There are two types of modules: motion units, which use a hinge moved by an electric motor to inch along the floor, and node units, which have multiple attachment points so that other modules can branch off in different directions. Each module houses a powerful computer-processing chip with a lot of memory, giving the robot an intelligence that distributes instructions and data throughout its structure, including commands that synchronize the motion hinges so that the entire PolyBot moves.

With its many motors, sensors, and computer chips, Polybot requires a good deal of electrical power, and battery life is a problem. Also, programming PolyBot becomes a struggle, according to the researchers, when many modules (units have been made with up to 100 modules) are involved. Nevertheless, the robot successfully demonstrates different styles of locomotion, including earthwormlike (the modules alternately squeeze and stretch), Slinky-like (end-over-end, like the child's toy), and caterpillarlike (many small feet). A PolyBot can also reconfigure itself, and not only for motion. It can create armlike limbs to deal with small objects.

A video designed to display PolyBot's versatility shows it making its way through a military-style obstacle course while a bemused U.S. Marine watches (the military connection is that the project is partly funded by the Department of Defense through DARPA). Another video, this one computer-simulated, shows PolyBot successively adopting different shapes and motions for different terrains: a rolling tank-tread-like loop for a flat surface, an earthworm to crawl down a step, and a multilegged spider for rough terrain. The configuration change from worm to spider is especially striking because it dramatically illustrates the effective operation of a robot utterly different from anything human.

ARMS, HANDS, FINGERS, AND THUMBS

Walking on two legs, as humans do, might seem a disadvantage compared to the flexibility of a PolyBot or to the stability enjoyed by four-legged creatures. But walking on two legs rather than four frees our arms, hands, and fingers to carry out the complex functions of grasping, holding, and feeling. This extraordinary versatility is not only an essential part of being human; it is one reason we are human, because our flexibility in manipulating the world has improved our thinking capacities.

That flexibility comes largely from the opposability of the human thumb—that is, it can be brought into contact with the tips of any of the four fingers. Although some other primates can also do this, the crucial point is that humans have a large area of contact between the thumb and the sensitive skin of the fingertips. This gives enormous dexterity in dealing with objects and their textures, which has strongly affected the development of our species. Benjamin Franklin was among the first to realize that the ability to manipulate is essential to human culture; it was he who defined humanity as "the tool-making animal." The physician and primatologist John Napier enlarges on this comment in his book *Hands*, explaining that the opposable thumb

> promoted the adoption of the upright posture and bipedal walking, tool-using and tool-making that, in turn, led to enlargement of the brain through a positive feed-back mechanism. In this sense it was probably the single most crucial adaptation in our evolutionary history.

But the power of a hand cannot be brought to bear without immense flexibility from shoulder to wrist, which in humans draws on six degrees of freedom. Two of these degrees represent rotation at the shoulder, raising the arm higher or lower, and rotating it backward and forward. A third is found at the elbow joint, and three more operate at the wrist—rotation around the axis of the forearm, movement of the hand up or down, and movement left and right. Designing so versatile a jointed system for an artificial being is an engineering challenge, and designing it to move properly is a geometric and computational one, as complex as programming a robot to walk on two legs, and sometimes requiring advanced mathematics. But the humanoid robots H6, H7, P3, and ASIMO all have arms and wrists as multijointed as human ones and are programmed to carry out some manipulations. H7 can reach under a table and grasp an object on the floor, and P3 can turn a handle to open a door.

However, what passes for hands on these units are poor substitutes for human hands. Rather than flexible thumb and fingers, they have a gripper design that grasps an object between stiff tines, with a squeezing action like a big pair of tweezers. Moreover, the robotic hands lack any sense of the weight or texture of an object. This limited design appears in prosthetic hands as well; Paolo Dario and his colleagues of the Mitech Laboratory, Scuola Superiore Sant'Anna, in Pisa, Italy, note that "current prosthetic hands are simple grippers with one or two degrees of freedom, which barely restores the capability of the thumb–index [finger] pinch."

Nevertheless, even limited capability is worth a great deal in a prosthetic replacement and is not necessarily a problem for many robotic applications. An assembly-line robot that manipulates automobile parts needs brute strength, not sensitivity, and works perfectly well with powerful arms terminating in pincers rather than fingers and thumb. Universally useful artificial beings, however, need functioning fingers-and-thumb hands that include appropriate sensory feedback. A differently shaped or unfeeling hand could not operate electrical switches and valves, or properly use screwdrivers and hammers, all of which are designed for the human hand.

A more profound issue is summarized in John Napier's comment "a lively hand is the product of a lively mind. . . . When the brain is empty, the hands are still." The inverse might also be true—limited artificial hands might ensure a limited artificial mind because the hands extend the abstract power of the brain into the real world. Even if the brain is made of processed silicon rather than living neurons, the addition of hands turns an isolated artificial intelligence into an interactive artificial being. Such a being could be designed to grow in knowledge and capacity as it explores the world with its questing fingers, just as the human brain has grown through a two-way feedback with the body's own searching fingers.

Writing in the late 1970s, Napier expressed little hope that an artificial hand could be built. "There is nothing comparable to the human hand outside nature," he said, "for all our electronic and mechanical wizardry, we cannot reproduce an artificial forefinger that can feel as well as beckon." Although the human hand is indeed difficult to copy, since Napier's pessimistic statement, we have come a long way toward building robotic and prosthetic hands that work like natural ones. One leading example is under development at NASA's Johnson Space Flight Center in Houston, in conjunction with DARPA. NASA has considerable expertise in building wheeled robots, like the Sojourner rover that explored the Martian surface during the 1997 Pathfinder mission, but the artificial hand is different. It is the critical part of Robonaut, a robotic astronaut designed to stand in for a human when work must be carried out in space on Earth satellites, or on the International Space Station now taking shape in orbit.

Astronauts cannot breathe in a vacuum and need considerable time to suit up before venturing into space, time that could be critical in an emergency,The nonbreathing Robonaut can be ready to go at a moment's notice while the astronaut remains safely inside the spacecraft. Because Robonaut is designed to work in space where it will float weightlessly, it does not need a full body but might eventually be outfitted with a single support leg. At present it consists simply of a torso with attached arms. Covered in a protective white synthetic material and surmounted by a sleek helmet, the robot resembles nothing so much as an Olympic fencer clad in the classic white jacket and protective headgear.

Robonaut is not autonomous. It is remotely controlled by a human who sees three-dimensionally through two video cameras in its head that send their images to the operator's eyes through special goggles. The operator's hand is inserted into a data glove that faithfully converts movements into electronic signals, which the robotic hand obeys to mimic the operator's hand movements. The artificial hand matches a natural one in size and shape, but it is less flexible. It has 12 degrees of freedom, and two more in the up–down and side-to-side movements of the wrist (all operated by electric motors), compared to the 22 degrees of freedom in the human hand. Robonaut's thumb is opposable only to the pointing and index fingers, and this triad is used where dexterity is needed, whereas the remaining two fingers bend back and forth toward the palm to aid in grasping. While this is nominally a limitation, Robonaut's designers note that it is still more dexterous than a human astronaut constrained by a spacesuit.

Operating the robot is an immersive experience that Robonaut's creators call "tele-presence" even without feedback of force or tactile information from the hand (which is now under development). The operator tends to sense the robot's body as his own, illustrated in one incident where the operator moved his feet when Robonaut dropped a tool, although the robot has no feet, and so the human's own well-developed bodily intuitions help him control the remote hand. Robonaut has proven adept at handling an electric drill, giving injections for medical emergencies, working with hardware components used in the International Space Station, and, in one demonstration with implications for antiterrorism operations, opening a backpack, sorting through its contents, and choosing and removing a particular item—demonstrating how Robonaut might be used to search for a weapon or explosives while its operator remains at a safe distance.

With its human-appearing outline but different joint construction and texture, Robonaut's hand combines organic and machine looks, and there is a cyborglike aspect to the entire robot. The operator, although in no way neurally plugged into the robot, functions as the partly merged cognitive center of a piece of machinery, like a brain transplanted into a mechanical body. Robonaut is not the only remotely operated robot. The Utah-based Sarcos Corporation, for instance, has demonstrated a humanoid, human-size unit whose arms and legs follow the actions of a human operator (plugged into what amounts to a full-body data glove) well enough to dance, although not very gracefully.

Robonaut and the Sarcos robot represent halting first steps toward beings as capable as the fictional RoboCop and Deirdre. But researchers are getting closer to those sophisticated imaginary cyborgs, driven primarily by the desire to make better prosthetic devices, with new tools to accomplish that. Gerald Loeb and Frances Richmond, of the University of Southern California, note that although two centuries have elapsed since Galvani observed that electricity makes muscles twitch, only in the last three decades have roboticists and neurophysiologists begun seriously addressing the problem of how to make an artificial limb move under neural control. Now, they say, we have reached the point where it "appears feasible to graft robotic and electrophysiological instrumentation onto a biological system to repair it," but also note that this requires

> many channels of data transmission in each direction. These channels must be installed and function safely and reliably in one of the most challenging environments conceivable—the human body.

The possibilities, and the problems, for direct neural control of limbs are illustrated by a unit proposed by Paolo Dario and his colleagues. This effort aims at a prosthetic or "biomechatronic" hand designed to replace a missing hand with one that looks real and copies natural motion and sensing abilities. This design also lends itself to bionic enhancements, as well as use in artificial beings—a prime example of the fruitful crossovers between medical applications and robotics.

Like Robonaut, the biomechatronic hand gives priority to the thumb and first two fingers, which move around an object and fit themselves to its shape to grasp it. The three digits are driven by electric motors and linkages tiny enough to be embedded in the palm and the fingers without making the hand look unnatural; however, the remaining two fingers are not motor-driven. Dario's group has not yet built an entire hand, but they have made a prototype plastic finger, with two degrees of freedom corresponding to the two joints of a human finger. Despite the smallness of the motors in the artificial finger, the force it exerts as it bends is comparable to that from a real finger, and so the artificial hand seems capable of fine manipulation at least.

As far as its mechanical design goes, the biomechatronic hand in its present form could serve as a robotic appendage. But Dario's project also includes efforts to link the mechanical device to the human nervous system to make an advanced prosthesis, and so enters the area of "neurorobotics." As defined by John Chapin of the State University of New York and Karen Moxon of Drexel University, " 'neurorobotics' seeks to obtain motor command signals from the brain and transform them into electronic signals suitable for controlling a robotic device." Those commands could come from the cortex of the brain or from the peripheral nervous system.

The cortex is the thin, wrinkled layer of "gray matter" that covers much of the brain. It performs higher intellectual functions and deals with sensory information, speech, and motor activities. The peripheral nervous system is the part of the nerve network that carries outside stimuli to the brain and returns appropriate responses. Even when a hand or limb has been amputated, the neural signals that once controlled the appendage are still generated in the cortex and sent to the appropriate nerves. The goal is to extract these neural pulses at the cortex or in the peripheral system with an interface that converts them into electrical signals that can control a prosthetic hand (or any device), and, working in the opposite direction, that accepts electrical pulses from the hand (for instance from tactile sensors) and converts them into meaningful neural signals for the brain.

The technique explored by Dario's group is one of several used or proposed for such electroneural interfacing and relies on an interface implanted into the peripheral nervous system. This approach uses a so-called regeneration type of neural interface, which blends techniques from electrophysiology and from the well-established nanotechnology used to manufacture silicon computer chips. Silicon is a good material for implantation because it seems to be nontoxic and has the necessary mechanical and electrical characteristics. Also, it can be readily manipulated by using state-of-the-art chip technology at the small scale of the human nerves.

The interface unit begins with a minute square of silicon 1.5 millimeters (0.06 inches) on a side. A set of metal electrodes is deposited on the surface of the chip, and further processing produces an array of tiny square holes, measuring only thousandths of an inch across, that pierce the chip. Then the whole chip, now called a die, is placed within a small conduit made of a nontoxic plastic that will not deteriorate inside a living body. The electroneural connection is accomplished by cutting a specific nerve and letting it regenerate within the plastic conduit. Nerve fibers reconnect themselves through the holes in the die, bringing them near the electrodes, so that current can flow in both directions between nerve and electrodes. Tests in which the interfaces were installed in rabbits showed that the interchange of current works: Signals originating in the nerve were detected in wires attached to the electrodes on the die, and external electrical signals sent to the electrodes affected the nerve, making the rabbit's leg twitch.

For all the preliminary success of the rabbit experiments, serious obstacles stand in the way of extending the technique to humans. The neural signals are small and full of extraneous noise. They require complex processing to interpret correctly, and it has not yet been shown that the signals could actually control an artificial hand. On the biomedical side, we don't know if there would be harmful long-term effects on the body from the implants or if the combination of regenerated nerve and silicon chip will hold up over a long period. And most important, cutting a nerve is not something to be done lightly. Such a risk might be acceptable for an amputee who wants to replace a lost limb with an artificial one, but it might not be so for an uninjured person seeking only to be bionically enhanced.

Nevertheless, setting risks aside for the moment, in principle the regeneration-type interface dramatically illustrates how neural science combined with nanotechnology enables the transmission of internal neural signals to and from external digital electronics. The intrusive nature of this approach is an issue, but less invasive methods, which I will describe in Chapter 8, might be more acceptable. Electroneural interfaces have also made it possible to create living–nonliving hybrids where the living part consists of neurons or low-level animal systems, also considered in Chapter 8.

SMOOTHING THE MOTION

While the successful merging of the biological with the artificial is still under development, in one respect artificial beings have clear biological roots: In gross outline, artificial beings largely resemble natural living beings—if not humans, then animals such as snakes. Even the formless multiunit PolyBot moves somewhat like an earthworm or spider. The reason is easy to see. Many contemporary roboticists would agree with the engineer in Karel Capek's *R.U.R.* who says, "[T]he product of an engineer is technically at a higher pitch of perfection than a product of Nature. . . . God hasn't the slightest notion of modern engineering." Nevertheless, evolution has produced, in the billions of years of earthly life, varied and workable solutions to the problems of motion and manipulation, from insect legs and mammalian limbs to the ribs of the snake and the tentacles of the squid.

This does not mean that roboticists cannot find solutions that go beyond what nature provides. There are natural limits to the strength and speed of animals. But some natural solutions, such as the extremely rapid skittering motion of a cockroach, offer unique possibilities for artificial beings. However, nature does not move by means of wheels, gears, or servos, and so artificial beings do not move like animals.

Roboticists sometimes deliberately copy animal locomotion, as in Shigeo Hirose's snakelike ACM R-1 unit. Examples like that reflect a general perception that if we study biological systems, we can use the principles we learn to improve our own creations. Applied to locomotion, this biomimetic approach draws on biomechanics, the science of how animals move. One of its leading practitioners is Robert Full of the University of California at Berkeley. The name of his Poly–P.E.D.A.L. Laboratory reflects what he does, which is to study the performance, energetics, and dynamics of animal locomotion in multifooted creatures.

Full's main goal is a biological one: to understand the mechanical performance of animals in quantitative terms, building toward a general theory of how animal bodies have developed and how they work. The significance of this is to better grasp the enormous diversity we observe in animals and relate it to their evolutionary history and ecological significance. But as Full learns in detail how animals walk, run, climb, jump, and bounce around the world, his findings provide remarkable inspirations for the designers of robots. Full's studies of lizards, centipedes, and other creatures have contributed to the design of, for instance, legged robots that maintain balance with fewer cognitive resources than conventional designs and a crablike unit that can move on land or underwater, and in creating Mecho-Gecko, a robot based on studies of gecko lizards and cockroaches that can climb walls.

Another aspect of biomimetics is to understand how to build machines that look more human or animal-like, more natural, as they move. A walking robot, even one with many degrees of freedom in its legs, simply does not walk like a person. Our limbs move quietly and smoothly as our muscles contract and extend, whereas a robot's movements are defined by rigid mechanical linkages driven by precision servos. Hence a robot's motion does not look quite right when it performs large-scale movements such as are involved in walking, or fine movements such as the changes in eyes and mouth that animate human faces.

Kismet, the robot built by Cynthia Breazeal at MIT to explore human–robot interactions, shows to great effect the importance of small-scale movements. Much of Kismet's impact on people comes from the humanlike motion of its head and face. Kismet, says Breazeal, "could actually make eye contact with you. It's night and day when something looks into your eyes versus at your face or just at you. Eye contact is profound." Kismet's head movement and changes in expression are accomplished by approximately two dozen servos. They drive, for instance, its lips, which are sufficiently flexible to show different expressions, but are not meant to reproduce the versatility of natural human lips.

What is important about Kismet is its demonstration of the power of a social component in human–robot relationships, even though the robot is not a human replica. The social connection might be enhanced, however, with robots whose facial and bodily movements appear natural, and this objective might call for a different system than motors and mechanical linkages. An alternate approach that shows promise for simulating natural movement comes from new classes of "smart materials," which change their properties according to external stimuli or environmental conditions. For example, the plastics called electro-active polymers (EAP) change length or shape under electrical voltage, mimicking what natural muscles do under neural control. EAP materials date to the late nineteenth century, but they came under serious study only in the 1990s. In one type, electrons in the material set up an electric field that causes stretching or shrinkage. In another, which usually operates in a liquid environment, the motion of charged atoms—ions—makes the material bend. According to the Jet Propulsion Laboratory's Yoseph Bar-Cohen, a leading researcher in the area, "the main attractive characteristic of EAP [materials] is their operational similarity to biological muscles, particularly their resilience."

Along with Bar-Cohen, researchers around the world are exploring these materials. One dramatic example of these efforts can be seen in action at the Artificial Muscle Research Institute (AMRI) at the University of New Mexico, where a skeleton called Myster Bony has been fitted with artificial leg muscles. Perched on a fixed exercise bike, with its muscles hooked up to electrical power, Myster Bony pedals away indefinitely—or at least as long as the electricity holds out. Another example at AMRI is an artificial fish slowly swimming through the water as its tail, powered by an artificial muscle, switches gently from side to side. AMRI, where researchers study several types of artificial muscles, involves both engineers and medical researchers; some of its projects provide hope for eventually using synthetic muscles to help people with muscular dystrophy.

Further uses of EAP and similar materials can be seen in Bar-Cohen's laboratory in Pasadena, every corner of which contains a different device that illustrates the versatility of smart materials. For example, NASA is constantly seeking reliable, lightweight materials for use in space, so Bar-Cohen designed a tiny EAP-driven windshield wiper that sweeps back and forth to dust off a small glass observation window. As Bar-Cohen puts it, this is a device where "suddenly the material is everything," meaning it performs without conventional moving parts such as gears and bearings, a tremendous advantage for space applications. NASA approved the wiper for use aboard a small, wheeled robot called Nanorover. With a mass of 1.1 kilograms (2.5 pounds), Nanorover was designed to explore and send back data from a small asteroid, much as the Sojourner rover explored the surface of Mars.

Also in Bar-Cohen's laboratory is a model of a human head sculpted by David Hanson, who has designed animatronic entertainment robots for the Walt Disney Company. Its facial details, texture, and coloring are persuasive, down to the slightly bloodshot eyes: but the most fascinating feature is that its eyes and mouth are moved by artificial muscles of Hanson's own design. Although these muscles do not employ EAP materials, they show what synthetic muscles can do: With the power turned on, the mouth smiles and the eyes move (one winks), a powerful example of the power of muscles over gears in robot animation.

Bar-Cohen, along with Cynthia Breazeal, has edited a book called *Biologically-Inspired Intelligent Robots* to further explore the possibilities for artificial muscles. He notes, however, that artificial muscles have a long way to go to become effective in robots. For instance, the relatively weak forces that EAP materials exert limit the strength of an artificial limb and need to be enhanced. He also admits to some doubts about the creation of highly advanced artificial beings. "I am concerned," he says, "because once you release a technology you never know which way it is going to be developed."

But he feels entirely confident about designing prosthetic devices that use artificial muscles, and holds out some dazzling speculations for the future. One type of material now under study can change both its configuration and its color under electrical stimulus, which might lead to artificial faces that not only smile and frown but also blush. And Bar-Cohen imagines putting EAP materials into a form that can be sprayed out from a special printer much as ink droplets emerge from inkjet printers. The result might be an EAP-operated butterfly printed flat onto a sheet of paper, ready for shipping, that flaps its wings and flies off when released from its box.

It's a far cry from the ponderous walk of Gort the robot to ASIMO's confident stride and on to the smaller motions of Kismet's expressive face. These movements are important in the usefulness and acceptance of artificial beings, with small motions as meaningful as large when it comes to eliciting human reactions.

Successful artificial bodies, however, require more than just the right facial expression. Witness Bar-Cohen's arm-wrestling challenge. In 1999, to stimulate researchers, he set them the task of building an arm driven by artificial muscles that could defeat a human arm wrestler. Although this challenge has yet to be met, it is significant as a kind of Turing test for machine physicality. Turing's original test for machine intelligence depended on verbal ability, to be judged by a machine mind's response to queries. Isaac Asimov's story "Robbie" presents a kind of Turing test for machine feelings, as little Gloria judges her robot to be as kind and patient as any human, through its responses to her needs. Bar-Cohen's challenge adds another Turing-like dimension, for we might ask a person to wrestle an arm protruding through a curtain without seeing what the arm is attached to. After wrestling, and perhaps losing, we ask the human to judge whether the arm's owner is human or artificial—and not only by the strength of the arm, but also by the strategy and tactics employed by the brain controlling the arm.

Building persuasively human artificial limbs and bodies is more than a matter of mechatronics or the faithful modeling of the human form. If an artificial arm is to do well at arm-wrestling, it must include sensory abilities that determine what the opposing arm is doing, such as judging the strength and direction of the forces it exerts, and it must draw on cognitive abilities to interpret that information and work out effective defensive and offensive movements. Just as a chess-playing computer has to weigh the moves made by its human opponent and develop an opposing mental strategy, an arm-wrestling artificial being must assess its opponent's physical behavior and work out a physical strategy. This kind of intelligent response requires artificial sensing and artificial thinking, the subjects of the next chapters.

7

The Five Senses, and Beyond

We apprehend the world and each other through our senses; without them, we could think, perhaps, but we could not deal with physical reality or engage one another. Similarly, an artificial being needs more than a silicon brain, more than metal limbs and plastic muscles. As a creature in motion, it must understand its environment in order to move freely and intelligently. To deal with humans, it must respond to their presence and communicate with them. These functions require sensory apparatus, backed up by cognitive facilities that interpret what is sensed and make intelligent decisions about interacting with the world.

Humans make such decisions based on vision, hearing, touch, taste, and smell. (Broadly defined, touch includes the tactile sense of pressure, along with sensitivity to heat, cold, and pain, as well as the kinesthetic senses that track the position of the limbs, bodily posture, and balance. These are often clustered together as the haptic senses, from a Greek root meaning "to touch.") Each of these human senses has an artificial counterpart but a digital creature can be effective without the full set, although a true android would need all five. On the other hand, artificial beings might employ senses humans lack, such as batlike sonar "vision" and sensitivity to radio waves.

We can hardly imagine an artificial being without some form of vision, which is deeply embedded in us. Much of the human cortex is devoted to visual cognition, far more than to any other sensory mode. Vision is our most effective means of exploring our surroundings, from detailed closeups to distant panoramas, and, through our superb ability to recognize faces and their expressions, it is critical for social interaction. (People who suffer from the neurological condition called prosopagnosia, the inability to recognize faces, lead difficult lives. One sufferer tells of failing to identify his own mother, who never forgave him.) At a more abstract level, vision is an element in creating mental imagery, because the "mind's eye" uses some of the same mental facilities that carry out visual cognition.

We consider hearing to be our second most important sense. Like vision, it provides us with information about our surroundings, although to a lesser extent than in many animals. Working hand-in-glove with the power of speech, it is an important part of human communication, and although many animals use sound to communicate, language is a preeminent human ability—along with vision, one of our highest mental functions. Just as the act of seeing goes beyond the mere reception of light waves and attaches meaning to the images the waves form, meaningful speaking and listening go beyond the mere production and reception of sound waves.

Touch, taste, and smell require less mental processing than vision and hearing, and they engage the world more directly. With vision and hearing, we receive only energy; nothing material enters the body. Taste and smell, however, are the chemical senses that react to molecules actually penetrating the body. Tactile sensors in the skin also physically contact reality, determining what is hard or soft, hot or cold, enabling the hands to actively grasp and shape objects, and providing the emotional warmth of the human touch.

Emulating vision, hearing and speech, and haptic abilities would go far toward producing an effective artificial creature—possibly one that could develop further through its embodiment, as Rodney Brooks has proposed. This program omits smell and taste, which are essential for many living beings, as in the exquisite sense of smell in dogs, or the constant sampling of water by certain fish whose skin is covered by taste buds. Many animal species use pheromones, substances that transmit information from one creature to another by odor. Smell and taste do not play similarly important roles for humans, so these senses might seem like mere frills for artificial beings.

Nevertheless, artificial smell is important for uses such as detecting contaminants in air or water and can take on additional meaning because the sense of smell is linked to the fabric of thought. The human olfactory system has complex neural pathways, some going to the limbic system of the brain. This is a collection of interacting parts that appeared early in the evolution of the mammalian brain and is strongly tied to instincts and feelings. Odors can be powerfully evocative because they speak directly to this ancient core. This might seem irrelevant to machine thought, which we tend to characterize as rational rather than emotional. But a variety of evidence shows that reason and emotion are connected in our own brains and minds, as I will discuss in Chapter 8. True artificial thought might also require both and might be enriched by a layer of nonrational but valid meaning entering the brain through the sense of smell.

For now, though, artificial taste and smell are at an early stage where sensors are still being developed. This is also partly true for touch. However, we already have digital hardware that can detect and manipulate light, and sense and produce sound. Progress in artificial hearing, speech, and vision focuses on the cognitive abilities that support these three vital functions.

SEEING INTO KNOWING

Creating synthetic vision as powerful as the natural version is not easy, partly because the human eye is a remarkable optical instrument, with high resolution, the ability to distinguish millions of colors, and a variable focal length. But these features are enormously enhanced by the mind. Under mental control (largely at an unconscious level) your eyes automatically refocus to provide clear vision from near to far, and they constantly move, to ensure that the portion of the retina with the highest resolution points at the most significant part of a scene.

It takes further mental effort to interpret the information these actions bring into the brain. The brain must learn to see, a complex process that begins early in life. How difficult this is even for the powerful visual cortex is illustrated by a real-life case related by the neurologist and writer Oliver Sacks—the tale of a middle-aged man who miraculously regained his sight after decades of blindness, but who found that eyesight alone was not enough; he also needed a brain that had learned to understand visual information. Although he struggled hard to comprehend the world visually, it was too late for him to master this ability.

Given the enormous demands vision places on the brain, it is not surprising that it takes massive computing capacity for a machine to match human vision. Hans Moravec notes that early AI researchers were ready to believe that given the right software, machine minds could be made fully intelligent. "Computer vision convinced me otherwise," he now writes, adding,

> Each robot's-eye glimpse results in a million-point mosaic. Touching every point took our computer seconds, finding a few extended patterns consumed minutes, and full stereoscopic matching of the view from two eyes needed hours. Human vision does vastly more every tenth of a second.

Typically, to perform the equivalent of human vision in real time requires a computer executing billions of instructions per second. Early computers were incapable of handling streams of visual data and interpreting it on reasonable time scales; in the late 1960s and early 1970s, it took hours for the pioneering robot Shakey to calculate its actions as it scanned its surroundings.

Now cheap, readily available microprocessors can handle visual information at high speeds, and a laptop computer can perform aspects of visual cognition in real time. Larry Matthies, who runs the Machine Vision group at the Jet Propulsion Laboratory, says that computers are now so fast that even complex programs for machine vision can be rapidly executed. Philosophical differences about top-down versus bottom-up or other approaches to artificial vision, he adds, have "very quickly become outdated. Because we've got fast enough machines you can do better vision, more reasoning—and that's the solution."

Video cameras are the eyes of these fast processors, capturing images in digital form; that is, as streams of bits representing the position and color of each picture element or "pixel" in a video frame. A pixel is the smallest unit in an electronic display. It takes about a million pixels to form an image on a computer screen, just as a myriad of individual colored tiles forms a wall mosaic. (To be exact, computer monitors typically display 1,024 × 768 or 1,280 × 1,024 pixels horizontally and vertically, respectively). Even fewer pixels per frame is adequate for many uses, and that lower resolution is easily achieved with inexpensive Web cameras that routinely send video over the Internet.

Other approaches work differently from the eyes; they examine the environment actively rather than passively. One method employs low-power infrared lasers mounted on the artificial being. When the laser beams strike an object, they are reflected back to sensors mounted on the being, where their time of flight is analyzed to find the object's range and bearing. Another approach emulates the echolocation used by bats and porpoises. These creatures generate high-frequency (ultrasonic) sound waves and listen for the echoes, which their brains analyze to characterize their surroundings. A similar process operates in sonar (sound navigation ranging) as used by nuclear submarines, and some robots use sonar as well.

There are also new ways to interpret sensory data, such as the promising approach called probabilistic robotics. According to Sebastian Thrun (then at Carnegie Mellon University, and now at Stanford), it uses the fact that "robots are inherently uncertain about the state of their environments," because of limitations in their sensors, random noise, and the unpredictability of the environments themselves, caused by, for example, the movement of people within the creature's visual field. Instead of calculating exactly what to do next, the being accommodates its uncertainty by determining a range of possibilities. As its sensors gather more data, the being's calculations converge to a high level of confidence about its physical location and other quantities. This method takes more computer time than direct approaches, but today's computers are up to the task. As Thrun notes, the payoff is that a probabilistic robot can

> gracefully recover from errors, handle ambiguities, and integrate sensor data in a consistent way. Moreover, a probabilistic robot knows about its own ignorance—a key prerequisite of truly autonomous robots.

These sterling qualities sound like a working definition of mature human wisdom, and could provide a superior basis for a high level of robotic intelligence.

Despite these and other advances, no artificial being so far displays general visual comprehension at the human level, but artificial vision works well within certain categories essential for beings that are mobile or meant to interact with people.

FROM HERE TO THERE

To move from one location to another, an artificial being must know its starting position, plan a route, and make the journey without hitting anyone or anything—hence localization, mapping, and obstacle avoidance form a basic set of visual abilities. Using these abilities, more or less autonomous mobile digital beings are becoming almost common sights in a variety of arenas—the home, hospitals, museums, the battlefield, and on distant planets as part of NASA's exploration of space.

NASA cannot yet send astronauts to other planets, so the agency has pioneered in developing mobile robotic stand-ins for human explorers. The Robonaut unit described earlier is not one of these stand-ins, because the focus is on moving its arm and hand rather than its whole body, and its visual cognition comes from a human operator. But the Sojourner rover, a small, wheeled unit delivered to Mars by the Pathfinder mission and that began examining Martian rocks on July 4, 1997, was the first in a series of mobile exploration robots with visual abilities.

The latest NASA mission to Mars began with two spacecraft launched in June and July, 2003, each carrying a new rover. In January 2004, the spacecraft delivered these nearly identical robots—dubbed Spirit and Opportunity—to two widely separated areas of the planet, carefully chosen because they show signs that liquid water might have flowed there in the ancient past. If the robots determine that liquid water once existed on Mars, they will have found an important indicator for the existence of past Martian life.

Like Sojourner, Spirit and Opportunity carry instruments to examine rocks and soil, in the hope of finding detailed geological evidence for the past presence of water. However, the new rovers travel much faster than Sojourner did, covering in three Martian days the same 100 meters (330 feet) that Sojourner took 12 weeks to cover. An Earth-based controller can send a radio message to a rover telling it what to examine, but even at the speed of light, radio waves from Earth take minutes to reach Mars, making it impossible to drive the robot in real time. Thus an exploring rover is on its own and must see well enough to safely reach a specified site over rough terrain.

A rover does this by first determining its present location. It could do so by tracking every turn of its wheels since leaving its landing site, like an automobile odometer. However, wheels tend to slip on rocks and sand so instead the rover uses what Larry Matthies calls "visual odometry." Seeing the world in three dimensions through two video cameras, as we do through our eyes, it maps the peaks and valleys, the rough and smooth areas of its neighborhood. Then it selects a prominent benchmark feature, perhaps a tall rock with a distinctive shape that it can recognize from varied distances and angles. Referring to this landmark, the unit can determine where it is to within 1 percent of the distance it has traveled. After establishing its location, the rover plans its trek to the target area. Like a human mountain climber scanning the terrain ahead for the best route, it examines its three-dimensional map to determine surface roughness, grade steepness, and obstacles, and selects the best path.

If all goes as planned, this version of autonomous robot vision will play a central part in a mission costing $800 million. NASA sees the current mission as a prelude to a 2009 one, where an even more capable rover will move to selected rocks, pick them up, and carry them back to a spacecraft that will return those pieces of Mars to Earth.

Selecting visual landmarks for navigation also works on Earth. Paolo Pirjanian, Chief Scientist of California-based Evolution Robotics, Inc., sees the method as a boon for everyday use. Although robots are now used in applications such as delivering hospital supplies, they require training to familiarize them with their particular environment, possibly relying on effective but expensive laser rangefinders. Pirjanian and his colleagues propose an alternative they call visual simultaneous localization and mapping (VSLAM), which might be suitable for consumer products because it uses inexpensive video cameras.

A VSLAM robot gets its bearings in bootstrap fashion. It begins by taking pictures of recognizable features like furniture, and holds them in a database. Initially the robot estimates the landmarks' locations and its own through wheel odometry. As it continues mapping, it compares whatever its camera registers to its database. When a match occurs, the unit uses probabilistic methods to recalculate the landmarks' position and its own. The interplay between these upgrades steadily refines the robot's knowledge, leading to a final accuracy of about 10 centimeters (4 inches) in its position, and 5 degrees in its direction of motion. Unlike robots that find their way by means of a fixed internal map, VSLAM can also deal with change: If there is enough alteration in its surroundings that no landmarks are recognized, the robot finds new ones and updates its map.

Artificial vision has become so fine-tuned that it can be trusted at high speeds and when lives are at stake. The small robot cave explorers deployed in the 2001–2002 U.S. campaign in Afghanistan show the military potential, and the Department of Defense (DoD) foresees more demanding applications. Through DARPA, the DoD is offering $1 million to anyone who can create a self-guided unit for desert warfare. The prize will be awarded in 2004 for a vehicle that can trek through the Mojave Desert from Barstow, California to a location near Las Vegas on its own. To cover a distance of about 320 kilometers (200 miles) within the allotted time of 10 hours, the unit must maintain an average speed of at least 32 kilometers per hour (20 miles per hour).

Other high-speed applications aimed at improving automobile safety through the use of intelligent artificial vision have been under development at Carnegie Mellon University and elsewhere. In one current effort, the DaimlerChrysler Corporation is working on machine vision for its vehicles that would supplement and even override human judgment. Using video input, a fast computer in the vehicle keeps track of nearby objects in real time. "If a child suddenly appears between parked vehicles," says the corporation,

> the computer registers the danger within 80 milliseconds . . . and, if necessary, initiates the braking procedure. In this time the driver's visual center would only just have received the visual information . . . without the brain having been able to initiate any reaction at all.

This application reminds us that when artificial vision is not being used to examine other planets, it is operating in environments that include people. Whether to sense a child in traffic, or to enhance human–robot interactions in general, the ability to differentiate people from things is the next important level of artificial vision.

FACES IN THE CROWD

It's hard to imagine a more commonplace activity than recognizing a friend, but there is nothing simple about the action. His or her face must be detected as a face among many objects in the visual field, then recognized as belonging to a particular person. After that, we might also perceive the mood it is expressing. Human visual cognition is remarkably competent at all this, even with wide variations in lighting and in the angle at which we see the face, even if it is partly obscured or we have not seen it for a long time—so competent, in fact, that we sometimes see faces where none exist, as on the surface of the moon.

The realities of today's world provide strong motives to find ways of artificially replicating these abilities. With terrorism as a serious threat, with identity theft and transactional types of fraud growing, governments, law-enforcement agencies, and commercial enterprises seek secure and rapid means to verify personal identity. Computer methods can provide this service, within the area called biometrics—the identification and recognition of people through physiological or behavioral traits, which also includes fingerprinting, retinal scans, and voice recognition.

The same biometric capabilities that enhance security can also improve the interactions between artificial beings and humans. The first step is detecting that a face is present. One research group, led by Takeo Kanade of Carnegie Mellon University, has in the last several years found accurate ways to pick out faces from complex cluttered backgrounds, using probabilistic methods and also a neural net. As presented earlier, a neural net is a set of interconnected processors that can be trained to acquire and store knowledge—in this case, how to decide whether a given image contains a face. In the approach Kanade's group devised, the network examines a still image in small pieces, some chosen to filter for facelike features; for instance, one piece consists of horizontal stripes 20 pixels wide by five pixels high, a configuration that tends to pick out a mouth or pair of eyes in a face presented in full frontal view.

The researchers trained the network with more than a thousand assorted images of faces, and also with images deliberately chosen not to contain faces. As we ourselves do, the network sometimes incorrectly found faces where there were none. These erroneous choices became examples of what *not* to identify as a face, thereby sharpening the network's judgment. Once trained, the system was tested on hundreds of new images including photographs of individuals and groups, the *Mona Lisa*, and the face cards from a deck of playing cards. The network found up to 90 percent of the faces, depending on the tradeoff between making the identification highly certain and allowing a few incorrect identifications to slip through. The approach using probabilistic methods was even more effective, in that it also worked well for faces seen in profile and in three-quarters view. (You can try both approaches at Web sites maintained by Kanade's group, where anyone can submit test images. Each face that the algorithms find is returned neatly surrounded by a green outline, leaving no doubt of the effectiveness of the methods.)

This kind of face detection can also be carried out in real time, a requirement for robotic applications. One example of software for real-time detection, developed by the German-based Fraunhofer Institute for Integrated Circuits, can be downloaded from their Web site. Used on images generated by an inexpensive video camera connected to my desktop computer, this algorithm found a variety of real, photographed, and hand-drawn faces within a second of their appearance in the field of view, as long as the face was seen full on. The system could also track a face as it moved, if the movement was not too rapid.

The next step after detection, face *recognition,* is also reaching maturity, driven by pressing needs for identification and verification. In identification, an unknown face is compared to a dataset of known faces, such as a security watchlist; in verification, the claimant's face is compared to a stored image of the person he or she claims to be. Like face detection, recognition is susceptible to a variety of approaches, such as one developed by the MIT Media Lab's Alexander Pentland, who categorizes faces based on a set of visual building blocks he has developed; for instance, the appearance of the upper lip and the forehead. Computer software uses these fundamental elements to identify faces, with sufficient success that Pentland's method has earned the trust of banks and security agencies.

A recent series of tests of computerized face recognition systems that was sponsored by the FBI, the Secret Service, and other government agencies, proved that commercially available algorithms had significantly improved in just two years. Automatic verification software approved 90 percent of legitimate subjects and only 1 percent of imposters, and an unknown face was correctly identified as belonging to a base set of more than 37,000 faces, with virtual certainty or very high probability, more than 80 percent of the time.

Despite this impressive performance, the government tests showed that there are still kinks. Success rates dropped substantially when the subject was seen under some types of lighting. The rate of correct identification has also been low for faces not seen full on, but this problem has recently been largely alleviated by the "morphable model," in which the software generates a three-dimensional model of what the camera sees. This virtual face is then changed and rotated to show how the subject would look if facing forward, and the result is fed into the face recognition routine. In one example, a poor identification rate of 15 percent for subjects looking right or left jumped to 77 percent when the morphable model was employed. This improvement suggests that better software, coupled with increased computing capacity, will solve many if not all the remaining problems with face recognition technology.

If artificial beings are to "read" people; that is, read their emotions through their facial expressions, further advances are needed. The human face has more muscles than does the visage of any other living creature. These muscles can wrest the face into thousands of expressions, some differing only subtly but carrying serious differences in meaning. Since early studies made by the nineteenth-century anatomist Guillaume-Benjamin-Amand Duchenne, for instance, it has been known that the difference between a false smile of seeming happiness, and a true smile of real joy, is that in a true smile the corners of the mouth are raised and the skin crinkles at the corner of the eyes.

Machine vision can already distinguish among emotions that produce widely different expressions. In one example, Gwen Littlewort and her colleagues, at the Machine Perception Laboratory of the University of California, San Diego, have developed a system that automatically detects a face as seen in a video image, and decides in which of seven categories its expression belongs: anger, disgust, fear, joy, sadness, surprise, or neutrality. Although relatively crude, this level of emotional identification is sufficient to enhance rapport between humans and artificial beings, allowing the latter to respond differently to an angry person, say, than to a surprised one.

But a digital being that cannot tell a false smile from a real one might remain naïve about humans, like the android Commander Data in *Star Trek*. Fortunately, in 1982, Paul Ekman, a psychologist of the University of California, San Francisco, who specializes in facial expressions, with his colleague Wallance Friesen, developed a method to classify everything a face can do. The Facial Action Coding System uses anatomical knowledge to define more than 30 action units (AUs) corresponding to contractions of specific muscles in the upper and lower face. These AUs are sufficient to fully describe the thousands of possible facial expressions.

In 2001, Takeo Kanade's group at Carnegie Mellon drew on this work to develop a neural network that breaks down any facial expression it sees into discrete AUs, with a recognition rate exceeding 96 percent. This means only that the system can detect subtle differences in expressions, not necessarily the emotions behind them, but psychologists are working on associating specific emotions with specific combinations of AUs, so there is potential for artificial beings to be able to perceive the fine points of human feelings.

The techniques that work for detecting and recognizing faces, such as the probabilistic approach, can also be applied to objects like automobiles and paper money, so machine vision will grow in capability. What has been achieved so far is only a part of general human visual cognitive ability.

GETTING THE WORD OUT

Despite the long way left to go, though, recognizing people and reading their faces represents a landmark in the development of synthetic creatures. But to achieve a comfortable relation with people, an artificial being also requires intelligent hearing and speech. Both the virtual and the real histories of artificial beings recognize the power of meaningful discourse—from the brass talking head supposedly made by Albertus Magnus in the thirteenth century, to the Turing test. As noted earlier, in 1637 René Descartes asserted that it might be possible to construct a machine that uttered words. But he went on to say,

> It is not conceivable that such a machine should produce different arrangements of words so as to give an appropriately meaningful answer to whatever is said in its presence, as even the dullest of men can do.

We do not yet have machines that converse as well as "even the dullest of men": but we do have transducers that change sound waves into computer bits, and vice-versa. This is a start, and researchers have created systems that hear what is said to them and give appropriate spoken responses, but only within limited arenas. However, these machines also display qualities that Descartes might never have considered: They sound human, and in addition to grasping the meaning of the words, they grasp how the words are said and the qualities of the voice that says them.

It's easy to experience machine hearing and speech, at least at a rudimentary level. My I-Cybie robot dog has a microphone in each plastic ear to triangulate the source of a sound. When I clap my hands, the dog turns its head toward me. If I clap in a certain sequence or say one of a small vocabulary of command words, it does a trick, like any well-trained natural dog. Also like a real dog, it learns to respond to a name and it speaks dog language—barking to show that it understands a given command and whimpering when it is not getting enough attention.

Another level of speech interaction is found in computer dictation programs, where what you say into a microphone is turned into written words on the screen. To get a true sense of machine conversation, though, pick up the telephone and dial airline reservations or your bank. There is a good chance you'll hear a synthesized voice welcome you, and ask what you need. You respond verbally, and a dialogue ensues. The conversation might well have its moments of frustration when you and the machine misunderstand each other. Still, according to Julia Hirschberg, a computational linguist at Columbia University, such conversations represent significant progress since the late 1980s. Computers are now fast enough to hear and respond in real time, and although the process is not perfect, Hirschberg notes that "Speech recognition and understanding is '*good enough*' for limited, goal-directed interactions." (Italics in the original.)

To be judged good enough or better, a machine must pass three tests: It must recognize the words you say, regardless of accent and personal speaking style, it must generate words that you recognize without machinelike overtones, and it must give sensible responses to your conversation. This last requirement is basically the Turing test, only with speech instead of written messages. If the machine converses so well on any conceivable subject that it cannot be distinguished from a person, it passes Turing's criterion for artificial intelligence. Even in limited conversations, however, the computer must be able to recognize words spoken by people, and to form its own words.

Speech recognition systems work by matching what a person says against a corpus; that is, a dataset of natural speech stored in the computer. The bigger the corpus, the better the system can recognize a range of utterances. Each speech sound in the corpus is broken down into a soundprint or acoustic spectrum—a list of the frequencies that make up the sound and their strengths. When the system hears a voice, that, too, is analyzed, in real time. By comparing the incoming soundprints with the stored ones, the computer assigns a probability that each sound has been correctly recognized. Further information comes from knowing the probabilities of the myriad other sounds that might follow the recognized one. The system also uses a "dictionary," a set of sound prints for words in the language, and a "grammar," which tells it the probability of finding a particular word once the preceding word is known. Then all these factors are manipulated by extremely sophisticated statistics, resulting in highly accurate word recognition. Compared to this complex process, speech synthesis is relatively simple.

But merely recognizing and saying words is not enough. As researcher Sylvie Mozziconacci of Leiden University writes,

> Communication is not merely an exchange of words . . . variations in pitch, intensity, speech rate, rhythm and voice quality are available to speaker and listener in order to encode and decode the full spoken message.

Recognizing words is one thing. Interpreting them, or speaking them with natural meaning and delivery, is something else.

To make a synthetic voice sound better than the mechanical monotone of a movie robot requires prosody. To poets, prosody means the study of meter, alliteration, and rhyme scheme that contribute to the flow and impact of a poem. For those who design machines that speak and listen, prosody means the differences in intonation that people use in speech, adding meaning or emotion to the literal significance of the words, or, as Elizabeth Shriberg and Andreas Stolcke of SRI International write, it is "the rhythm and melody of speech." These intonational variations are put into synthesized voices by careful adjustment of pitch, pacing, and so on to copy the natural sound of people talking.

The other side of the prosody coin is the problem of ensuring that an artificial being can fully interpret what humans say. That helps to reduce the ambiguity in human language, a major barrier to full machine understanding of speech, and to sense emotions and physiological states expressed in the human voice. As in facial recognition, the aim of sensing emotions and physiological states is being driven by the war on terrorism because it is important to detect stress in intercepted voice communications or "chatter." Corporations are interested as well; they want to know when customers on the telephone are angry so that they can be mollified by appropriate responses (which might be hopeless if a customer's anger was elicited by the frustration of talking to a machine with poor verbal skills).

But it is not easy to quantify exactly what it is we sense in prosody, or to put that knowledge into artificial speech systems. Reviewing how humans recognize emotion, Ralph Adolphs, a neurologist at the University of Iowa, says,

> In general, recognizing emotions from prosody alone is more difficult than recognizing emotions from facial expressions. Certain emotions, such as disgust, can be recognized only very poorly from prosody.

Even human evaluators might disagree about how to classify the emotions expressed in a voice, especially for short utterances. Without a reliable database of classifications, it is difficult to determine exactly what a machine system should listen for to determine a person's state of mind. But progress is being made in this relatively new area, especially in its pragmatic aspects: for instance, it seems that when correcting an error made by an artificial speech system a human tends to hyperarticulate—that is, speak slower and louder, and at a higher pitch—a clue that is useful in helping the system to respond appropriately.

Today's artificial speech systems show the level at which recognition, synthesis, and conversational ability come together. Speech Experts, a German firm, recently announced a washing machine that obeys voice commands. This might seem an odd choice for advanced speech capabilities, but a company spokesman claims that, "Electronic appliances have become so complicated . . . that consumers are put off by them. Speech recognition would help people." The machine is said to be able to follow complex instructions, such as "Prewash, then hot wash at 95 degrees, then spin at 1,400 revolutions and start in half an hour." It currently responds to a few hundred German words, but is expected to be able to eventually handle several thousand, and in other languages as well.

Another example that shows how conversational machines function in practice is a telephone-based system for booking train travel, used since 2001 by Amtrak, the U.S. passenger rail system. Dial the Amtrak number, and a pleasantly crisp female voice says "Hi. This is Amtrak. I'm Julie." Speaking in the first person and using casual speech such as "Here goes" and "No problem," Julie offers schedules, ticket reservations, and train status. At each juncture where the caller must make a choice, the questions are crafted so that a yes or no will do, or Julie announces the words the customer can use and be understood, such as "Book that one" or "Change itinerary."

Within the constraint of a limited vocabulary, Julie does well in recognizing words and responding suitably, as I found when I decided to test Julie by making a reservation. In several conversations, it never missed "New Orleans," which has a variety of pronunciations. It misunderstood only when I departed from the list of approved words, and once when it interpreted my "19" as "90"—an understandable error that humans make too—and the system let me correct the error with little fuss. Surveys show that customers are substantially happier with Julie than with the touch-tone method Amtrak used previously—but the same surveys also show that many customers still hang up before completing the reservation process. Certainly no one yet has full confidence in Julie, competent as it sounds; the caller can always say "agent" to get connected to a human.

Other voice-based systems include a mock air-travel planning service based at Carnegie Mellon University that was designed as a test bed for the DARPA Communicator project. This ambitious effort had the goal of developing speech-based interfaces for battlefield use that would "support complex conversational interaction, where both user and the system can initiate interaction, provide information, ask for clarification, signal nonunderstanding, or interrupt the other participant." When you dial the phone number, you are greeted by a male voice. Its timbre is pleasant, but its delivery is a touch robotic, so although this system delivers the same kind of information Julie does, it is a less engaging chat partner.

The limitations of current voice systems are clear, and their variations emphasize that there is as yet no single optimum approach, although some methods are perfectly adequate for closely constrained dialogue. To achieve a higher standard of machine listening, understanding, and speaking that approaches human levels, deeper aspects of artificial intelligence must come into play, as Alan Turing understood. But even at the lower levels we have achieved so far, there is undeniable power in hearing a humanlike voice respond to your words—or perhaps hearing a digital being greet you by name after recognizing your face, while extending a hand to shake yours.

REACH OUT AND TOUCH

Appealing as the idea of shaking hands with a humanoid creature can be, you might want to think twice about actually doing it. Entrusting your fingers to a motor-driven mechanical hand could lead to pain or worse. An artificial being might know enough to begin grasping your hand in a socially acceptable way, but not when to stop. When you shake hands with another person, you each feel the pressure the other is exerting. Unless your intention is the hostile one of squeezing as hard as you can, you modulate your grip to more or less match what you feel from the other hand.

Artificial beings need a similar kind of sensing, and not only to keep from hurting humans. If a being can track the forces that it develops as it interacts with its environment, it can precisely calibrate how to grasp things, and it can adjust the forces it exerts so that its appendages "give" when they encounter an obstacle. This force feedback is one essential for artificial touch. Another is the kinesthetic sense that gives information about the location of a being's limbs; otherwise, it could not guide its own hand toward an object. A third is a true tactile sense, allowing the being to perceive the surface properties of whatever it manipulates.

These haptic modes are found in varying degrees in artificial creatures. Kinesthetic sensing is a necessity for the walking robots in Chapter 6, and units with hands need broader abilities. Two examples with cyborglike elements have been developed for use in space: NASA's Robonaut, and the four-fingered DLR Hand II, developed at the Deutschen Zentrum für Luft- und Raumfahrt (DLR), the German Aerospace Center. In both Robonaut and the DLR Hand II, human operators remotely perform manual tasks using video feeds that display what the robotic hand is doing. But the humans do better when the forces and textures felt by the robotic hand are fed back to their hands, via data gloves.

The transmission of sensory data from a robotic hand to a real one requires ingenious and extensive hardware. The force sensors developed for the DLR Hand II are tiny enough to fit into its fingertips, and according to Robert Ambrose, who heads the Robonaut group, the unit has more than 150 sensors in its arm and hand, although not all are involved in providing feedback. But even this many sensors is not enough to match the full power of human tactility. Our fingertips and tongue-tip are highly sensitive because touch sensors are densely concentrated there. We do not fully understand this network, and some researchers think its complexity rivals that of the visual system. In any case, it takes clever engineering to make sensors small and numerous enough to be installed at similar high densities.

The engineering challenge is being addressed, however, because of the role artificial touch can play in robotic surgery, a technique that is now commercially available, for instance, in Intuitive Surgical's da Vinci system. Like the NASA and DLR robots, surgical robots are cyborglike rather than autonomous; that is, a trained human surgeon manipulates controls to operate a remote set of surgical instruments. One day, surgeons might be able to operate remotely at accident or battlefield sites anywhere in the world. Another application is already realized—minimally invasive surgery, performed through small bodily incisions typically a centimeter in size. The surgeon sees by way of a tiny video camera called an endoscope, and wields miniature tools, all inserted through the incisions. With the intervention of suitable hardware and a computer, the surgeon's hand movements are appropriately scaled down, and any hand tremors are removed.

The technique is being used for a variety of procedures, from gallbladder removal to heart valve repair. It offers patients reduced pain and blood loss, minimal muscular damage, and shorter recovery times. However, one drawback is the surgeon's inability to directly feel internal organs and their resistance to the scalpel. To remedy this problem, force feedback and tactile sensing are being added to surgical robots, with encouraging results. At the Harvard BioRobotics Laboratory, Robert Howe and his colleagues monitored medical students as they used a telerobotic system to expose a simulated artery, a common type of surgical task. Adding force feedback to visual feedback did not improve the speed or precision of the operations, but it did enable the students to perform the procedures less forcefully. This reduced the rate of inadvertent damage or "nicking" of the artery by some 75 percent compared to remote surgery using visual feedback alone.

The Harvard group is also finding ways to help surgeons remotely search for internal lumps, not easy to do through a small incision. Howe likens it to "trying to find a pea inside a bowl of jello using chopsticks." The solution is a robotic fingertip consisting of 64 pressure sensors in a square array, inserted in the body. Each sensor is connected to a motorized pin outside the body, and the surgeon's finger rests against this array of pins. As the robot fingertip moves within the body and encounters a lump, the pressure readings on the sensors change and the corresponding pins move in proportion. The end result is that the array of external pins maps the shape of the lump, which can then be felt by the surgeon's finger resting on the pins.

Other approaches could eliminate separate sensors to yield artificial skin or muscles with built-in haptic senses. For example, researchers at the STMicroelectronics Corporation and the University of Bologna have mounted a grid of fine electrically conducting wires in a soft substrate. Pressure on the material changes the electrical interactions among the wires. This information is turned into a map that gives the shape of the object causing the deformation. And at the Polytechnic University of Cartagena, Toribio Otero and Maria Cortés have used a plastic called polypyrrole to make a touch-sensitive muscle. Like other smart materials used for artificial muscles, theirs alters its electrical properties in response to pressure and changes shape when an electrical current is applied. The interaction between these behaviors provides feedback that adjusts the force the material exerts according to the resistance it encounters, as we humans do.

Sensitive artificial touch is an engineering challenge because it requires many sensors that are densely distributed over an area; synthetic smell and taste are difficult to implement because of the sheer variety of what they sample. Nevertheless, concerns about security and crime are motivating researchers to develop artificial smell. A sensitive nose, natural or artificial, can detect explosives, buried land mines, and smoke from fires, as well as hidden drugs. Although the sense of smell is not fully understood, we know that humans identify smells by means of about a thousand special proteins in the nose, each of which reacts to a particular group of molecules, typically of an organic substance. Most odors do not come from just one chemical element or compound. When we recognize a smell as "coffee" or "vanilla," we are identifying a set of molecules that has activated a particular pattern of proteins, which means we can recognize many millions of odors.

An artificial nose, therefore, must first react to specific chemicals, and then register the different compounds in a given odor. Moreover, to become a useful digital technique, it must change chemical reactions into electronic impulses. The Cyrano 320, an electronic nose made by Cyrano Sciences of Pasadena, California, uses a small chip with 32 receptors. Each receptor consists of a specific polymer mixed with some carbon black, a form of carbon that conducts electricity. When exposed to a vapor, each polymer expands by an amount determined by the molecules making up the vapor. This expansion changes the electrical resistance of each polymer and hence of the entire chip, producing a composite fingerprint reflecting all the molecules the chip has detected. Although 32 receptors is not many compared to the thousand proteins in the human nose, it is still enough to identify a lot of odors.

An artificial tongue can operate in a similar way, because all the flavors we experience, from ice cream to sushi, arise when our taste buds respond to a basic palette: the traditional bitter, sour, sweet, and salty, with umami (the taste that comes with monosodium glutamate or MSG) recently added by many experts. The food and beverage industries have developed devices more sensitive than the human tongue to detect flavors, such as bitterness and sweetness, essential for their products. Researchers at the University of Texas and University of Connecticut have gone further, developing electronic methods to test for the presence of all the basic tastes except umami, although these methods have not yet yielded a commercial product.

MORE THAN HUMAN

Sight, hearing, touch, taste, and smell—for each, there are ways the artificial versions fall short of nature, but other ways they can improve on it. They can be extended beyond human norms, or supplemented by sensory modes without human analogues, such as active probing by sonar or laser beams, which work even in the dark, determine the distance and direction to an object, and distinguish between different types of obstacles.

Other advantages are realized by extending artificial vision further into the electromagnetic spectrum. Humans can see light from 400 to 750 nanometers in wavelength, from violet to red, with the other rainbow colors in between. This is only a tiny portion of the range for electromagnetic radiation, from X-rays and gamma rays with ultrashort wavelengths, to radio waves many meters in wavelength. Within this range lies invisible infrared radiation, which begins at wavelengths beyond 750 nanometers and is generally produced by objects hotter than room temperature. Hold your hand above a hot electric heating coil, or stand in bright sunlight; the warmth you feel is delivered by infrared waves.

The connection between heat and infrared radiation gives another way to see in the dark; that is, to discern warm or hot entities like human bodies and internal combustion engines. This is the principle behind one kind of night-vision goggle, and appropriate sensors provide the same capability to robots. The advantages for military, police, and rescue operations are obvious, and if nursebots or doctorbots ever become realities, their medical diagnoses could be aided by infrared vision. It can detect tumors, which are warmer than their bodily surroundings, and can remotely measure body temperature. This capability became important during the breakout of severe acute respiratory syndrome (SARS) in 2003, when international travelers were screened by testing them for above-normal temperatures that might indicate the high-fever characteristic of the disease.

Add radio waves to the suite of electromagnetic wavelengths that digital beings could sense, and you get another extrahuman mode. One result could be beings that always know exactly where they are. While a robot on Mars needs extraordinary means to determine its location, a unit on Earth could simply incorporate a global positioner—the small electronic device that uses radio signals from orbiting artificial satellites to determine where on the planet it sits, to an accuracy of a few meters. Artificial beings could also have complete access to the resources of the Internet, through high-speed wireless connections, giving them the ability to tap into a world of databases, factual information, news, and much more for the being's own use or to answer questions from humans.

With radio, artificial beings could also engage in artificial telepathy, silently communicating among themselves even when far apart. Recall the brutal worldwide uprising of robots in the play *R.U.R.*, or their sinister swarms in the story "With Folded Hands." It takes only a touch of paranoia to see robot telepathy as a threat, but the applications thus far have been benign. The best known such application is competitive soccer played by teams of wirelessly linked AIBO robot dogs. Robotic soccer has taught researchers a lot about coordinated robotic behavior, and it has also evolved into an annual World RoboCup event where crowds cheer on their teams, and wait for a player to score a goal and perform a victory dance. Similarly, a big hit of the ROBODEX 2003 exposition in Japan was a robotic ballet. The principal dancers and *corps de ballet* consisted of tiny inch-tall units, made by the Seiko Epson Corporation. Controlled by a wireless linkage, they gracefully twirled, blinked their LED eyes, and formed perfectly aligned patterns to the strains of romantic music, as audiences watched enthralled.

CHIP VISION

Artificial senses can also open up a whole world of new human capabilities; as bionic implants, they can not only replace but even extend the natural senses. There is enormous interest in doing for the blind what cochlear implants have done for the deaf, as well as in other possibilities for bionic enhancement or replacement of human sensory organs. A limited experiment in direct human access to wireless communication was carried out in 1998 by Kevin Warwick, at the University of Reading in the United Kingdom, who had implanted into his arm a chip that emitted an identifying radio signal. The signal triggered functions such as turning on lights when he entered a room. However, the chip was not connected to his nervous system and did not carry out any functions of greater complexity.

Now under way are substantial efforts to restore sight to the blind through implants in the brain or retina. Most blindness is caused by a loss in the retina's sensitivity to light, although both the optic nerve, which transmits visual impulses to the visual cortex, and the visual cortex itself remain perfectly functional. This is what happens to people with the disease called retinitis pigmentosa, and to those with macular degeneration—the age-related condition that is the most common cause of blindness in the United States, responsible for loss of sight in 200,000 eyes per year. In these cases, retinal implants show promise for restoring sight.

When a nonworking retina is electrically stimulated, the brain perceives flashes of light called phosphenes. Nanoelectronic techniques have made it possible to embed a minute set of electrodes, a fraction of a centimeter across, in the eye atop the retina. In one recent example, a group led by Mark Humayun and Eugene de Juan, at the University of Southern California in Los Angeles, implanted such an array connected to a video camera worn by the blind person. The camera activates the electrodes, stimulating neurons to create phosphenes that are related to the image registered by the camera. At the Illinois-based Optobionics Corporation, its founders Vincent and Alan Chow have eliminated the camera by implanting chips containing silicon light sensors directly into the eyes of test subjects. The sensors convert light into electrical impulses that activate nerve cells.

A more radical method brings visual information directly into the brain, which means the technique could cure blindness due to a damaged eye or optic nerve, as well as blindness arising from retinal problems. William Dobelle, an independent scientist who operates his own laboratories in the United States and Portugal, has developed an electrode array that is implanted on the surface of the brain, where it stimulates the visual cortex. The array is connected to an electrical socket mounted on the outer surface of the skull, into which is plugged a video camera.

None of the methods described above is a complete bionic cure for blindness. Many questions remain, such as how well the body accepts the implants. However, these initial efforts are providing glimmerings of vision to the blind—in one case, apparently sufficient to allow the implantee to drive a car under controlled conditions—although not yet anything close to full restoration of sight. One problem is low resolution, because the number of electrodes or sensors in each implant is minuscule compared to the millions of rods and cones in the natural retina. Advances in nanoelectronics will undoubtedly improve the resolution, but a more fundamental difficulty remains. The retina contains a complex multilayered system of neurons that respond to the impulses from the rods and cones and thereby analyze visual information even before it reaches the brain. This retinal processing tracks movement and the edges of objects, both significant elements in any visual scene. None of the implant schemes tested so far performs this essential first step in visual thinking, but this important point is being addressed by researchers working on "biomorphic" or "neuromorphic" chips that copy biological functioning.

Kwabena Boahen, at the University of Pennsylvania, has gone beyond merely simulating the retina to actually copying it. Using transistors etched in a silicon chip, which are interconnected and made to operate in a way that mimics the layered retinal neurons, he has reproduced the edge- and motion-detection carried out by a natural retina. There is still a long path ahead, however, before this chip is ready to be tested in a human subject. Indeed, there is a long path ahead until any of the retinal or brain implants can gain FDA approval, but the path might eventually lead beyond replacement to enhancement. Implants that use a video camera could draw on the advantages of telephoto, wide angle, and zoom lenses to enhance bionic vision. The camera could also be made sensitive to infrared light, giving the wearer night vision, which could also be built into implants that use light sensors in the eye rather than a camera.

Apart from implants, approaches like laser surgery combined with adaptive optics—the technique used in ground-based astronomical telescopes to correct light distortions caused by atmospheric turbulence—could bring us supernormal vision. The method relies on a wave-front sensor to examine the light waves; if they are not in perfect step, the deviations are corrected by changing the shape of a mirror as the light reflects from it, producing an undistorted image. David Williams and Junzhong Liang, of the University of Rochester, have pioneered the use of wave-front sensors to map all the optical aberrations in a person's eye. This technique provides guidance for an advanced form of laser surgery, where the surgeon sculpts the cornea with tiny compensating corrections. In principle, all vision problems including astigmatism can be fully eliminated to give the fortunate patient 20/10 or 20/8 vision—the absolute best the human eye can do, given its density of rods and cones. Clinical trials have shown the effectiveness of the technique, which has given some people 20/16 vision.

In both natural and artificial beings, the senses are bridges between the physical operations of a body and the higher operations of a brain or a mind. These sensory bridges carry us into the mental make-up of a digital being: its intelligence, its rational thought, its feelings if any, and—if any—its consciousness.

Thinking, Emotion, and Self-Awareness

Imagine an artificial being with a humanoid body and humanlike senses. Imagine it on its way to carry out a task, perhaps to retrieve a certain book from a certain desk. Watch it walk through its environment, adjusting its path so that it doesn't collide with people it encounters. Perhaps it recognizes someone, saying "Hello" and greeting the person by name; it might stop others and politely ask permission to scan their faces and ask their names, to hold in its memory. You might also see it giving reasonable answers to questions like "What's your name?" "Where are you going?" or even—as it consults its built-in wireless connection to the Internet—"Is the stock market up or down, and by how much?" When it finally reaches the right desk, it identifies the book by its color or title, picks it up, and walks back to where it started.

Each action, from walking to seeing to talking, has been described in earlier chapters as a separate achievement, although not necessarily all in one being or as part of a robotic body. Some artificial capabilities have meaning and value even if implemented only on a computer. Artificial vision and speech, for instance, have been goals of AI research apart from their use in robots because they are significant parts of human intelligence and because they have useful applications such as the translation of language by a machine.

But the creature we are imagining uses all these abilities to function in the world. Such sophisticated functioning requires integrated guidance and control of disparate body parts and actions—that is, it requires a brain. In humans, the brain is an enormous collection of neurons that controls our behavior along with the sensory and motor functions of our bodies. In an artificial being, the equivalent is an enormous collection of electronic switches—transistors—etched into silicon to make digital microprocessors and memory chips. This type of brain might be all that is needed to steadily make artificial beings more capable because the speed and capacity of silicon chips continue to grow. But other approaches are possible as well: different forms of electronic brains, brains that combine organic and electronic elements, perhaps even actual living brains inserted into cyborgs.

What a brain does is think, and what thinking imparts to a being, natural or artificial, is intelligence. Whether a being with an artificial brain is actually thinking while electrons course through its circuits is still a matter for debate. It is easier to ask, "Is the being intelligent?" because that question can be answered by observing whether it exhibits intelligent behavior. Alfred Binet, the French psychologist who in 1905 laid the groundwork for the intelligence quotient (IQ) test, defined human intelligence as the sum of mental processes that come into play in adapting the individual to the environment. Modern definitions agree that intelligence is an adaptive property, meaning that it helps the organism survive and thrive by providing effective responses to changing situations.

This general definition, based on the idea of effective responses to the environment, can also be applied to artificial beings. However, in his book *Behavior-Based Robotics,* Ronald Arkin takes the definition a step further to make it more explicit for robots. Borrowing his version and extending it slightly gives this working definition:

> An intelligent artificial being is a machine able to extract information from its environment and use knowledge about its world to move, and interact with people, in a safe, meaningful, and purposive manner.

Under this definition, a being that sees its surroundings and interprets them well enough to navigate, and recognizes people and their words well enough to respond to them, is exhibiting some degree of intelligent behavior.

We can refine our judgment of the intelligence of an artificial being through better understanding of human intelligence and its measurement. Contemporary researchers and educators tend to reject the traditional form of IQ test. They believe that it defines intelligence too stringently by weighing only linguistic and logical-mathematical abilities. Intelligence is now seen as a complex phenomenon with different facets that cannot be summed up by a single IQ number.

One view of this multiplicity comes from the pioneering work of the psychologist Robert Sternberg of Yale University. Sternberg has developed a theory of threefold intelligence that gives weight to analytical, creative, and practical components. Another view comes from Howard Gardner, Hobbs Professor in Cognition and Education at Harvard, who has written extensively about seven diverse types or categories of intelligence. These make a useful grid against which to measure artificial mental functioning.

Gardner begins with two categories that are cornerstones of most definitions of intelligence and adds others that might be less familiar. Paraphrasing Gardner's definitions, the seven are:

- Linguistic: Sensitivity to language, ability to use language to attain goals
- Logical-mathematical: Capacity for logical analysis and mathematical operations:
- Musical: Skill in recognizing and manipulating musical patterns
- Bodily-kinesthetic: Using one's body or parts of it to solve problems
- Spatial: Recognizing and manipulating the patterns of space
- Interpersonal: Understanding the intentions and desires of other people and so working effectively with them
- Intrapersonal: Understanding oneself, creating an internal working model of oneself and using it to manage one's own life

Where Gardner's intelligences do not exactly match some types of robotic capability, they involve them. Spatial intelligence, for instance—which Gardner describes as the kind of ability used by navigators, sculptors, and surgeons—relies heavily on visual cognition. In the same way, consciousness of self plays a role in intrapersonal intelligence. Some writers have combined the interpersonal and intrapersonal categories into "emotional intelligence," but it is useful to keep them separate when discussing whether artificial beings can be self-aware.

Gardner's original definitions include volitional and creative components such as "fashioning products" under bodily-kinesthetic intelligence. My paraphrases omit these components because artificial beings are not yet creative; nevertheless, digital beings do show traces of Gardner's intelligences. One of them, logical-mathematical ability, is essentially universal among artificial beings because each microprocessor chip has an arithmetic-logic unit (ALU) that accurately manipulates numbers at high speed. The ability to rapidly solve arithmetical problems and deal with areas, such as the stock market, that use arithmetic could be given to a digital being if it were a useful attribute. Artificial logical-mathematical ability could even extend to higher mathematics like algebra and calculus, using existing computer methods that manipulate mathematical ideas at an abstract symbolic level.

But some artificial creatures are more varied in their abilities. In fact, it seems that the seeds of each kind of intelligence can be found in existing beings, as we'll see by examining three especially intelligent ones. Two we have described before; the third closely parallels the capabilities of our imaginary unit.

THREE SMART DIGITAL BEINGS

ASIMO (advanced step in innovative mobility)—the walking robot made by the Honda Corporation, described in Chapter 6—is the oldest of the three smart beings. It took more than two dozen engineers some 14 years to produce this accomplished walking robot, which remains among the best. It displays a high level of kinesthetic intelligence. It can keep its balance as it walks on level ground or on a slope, and up and down steps. It can also balance on one leg, turn corners snappily, and walk backward.

Despite ASIMO's long lineage, though, there is plenty of room for improvement. At ROBODEX 2003, an older model striding along confidently at 1 kilometer per hour (0.6 miles per hour) was outpaced by a newer model that walked faster though less smoothly. With its flexible physical platform that boasts many degrees of freedom, and its humanoid bodily configuration, ASIMO is capable of even more. Honda states that its long-term goal is to make robots that "can be helpful to humans as well as be of practical use in society." At a height of 1.2 meters (48 inches), ASIMO can interact eye to eye with a person in a wheelchair or sitting up in bed, or with someone sitting at a desk. And with its childlike size, and its back-mounted computer that makes it resemble a student marching off to school with a backpack, ASIMO is quite the opposite of intimidating—it is endearing.

To extend its ability to deal with people, ASIMO's kinesthetic intelligence has grown, and new communicative—if not linguistic—abilities have been added. Although ASIMO does not yet speak very well, it can listen, react with its body, and read human body language. Call it by name, and it turns its head toward you. It can wave hello, and safely carry out the social gesture of shaking hands with a person. It comes to a halt when it sees a hand held upright, traffic-cop style, understands that a pointing human hand is a directive to go to the indicated location and wait, and recognizes a goodbye wave. Capabilities to identify faces and plot walking routes—that is, elements of interpersonal and spatial intelligence—have been added, and ASIMO can also carry objects in its hands. These capacities, integrated in a single robot, permit ASIMO to act as a receptionist that can greet and recognize visitors and guide them to a specified location.

The initial goal in developing ASIMO was to produce a walking robot; the other abilities were added later. In contrast, the robot Kismet was built to display what its maker calls social intelligence, or in Gardner's framework, interpersonal intelligence. Cynthia Breazeal, who, as a graduate student in Rodney Brooks's group at MIT, constructed the robot, is Kismet's guiding spirit. When she was a little girl, Breazeal writes, she was "captivated and fascinated" by two "compelling characters" from the 1977 science fiction film *Star Wars*, the robots R2-D2 and C-3PO. Now she describes her "dream of a sociable robot:"

> Taking R2-D2 and C-3PO as representative instances, a sociable robot is able to communicate and interact with us, understand and even relate to us, in a personal way. It is . . . socially intelligent in a human-like way. We interact with it as if it were a person, and ultimately as a friend.

Breazel argues for the importance of socially intelligent artificial beings, as they enter the human world with potential uses from entertainment to healthcare to military applications. Noting that we humans have become competent at social interaction because dealing with each other has been essential to developing a sophisticated human culture, she suggests that people relate better to technology that displays "rich social behavior." Socially intelligent beings could become colleagues with whom we communicate as easily as we do with people, and whom we even like having around. An important part of Breazeal's thinking is the idea that the right kind of interaction will encourage people to teach a being just as they would teach a human child, with important consequences for the future of artificial beings.

Social capability has been built into Kismet, which has no body, hands, or legs, only a head and face meant to interact with people through expressive movements and speech, with appropriate use of vision and hearing. As described earlier, the robot has exaggerated, clownlike features—big blue eyes, prominent lips, and conspicuous animal-like ears—which it moves to convey emotional states that people immediately grasp. Kismet does not look human, however. Breazeal chose not to attempt that illusion because an imperfect simulation of humanity can be disturbing. Instead, the design suggests a fantasy creature that acts believably human, an approach that has been brought to perfection in Walt Disney's animated offerings.

Further, Kismet's looks and behavior were chosen to give the impression that it is young. It has big eyes, and its visual system seeks primary colors and human skin tones and motion, so it reacts if a person shakes a brightly colored toy at it. Its auditory system listens both for words and, through prosodic analysis, the messages behind them, such as elements of praise, and it speaks in childlike tones. Judging by the responses of people who have chatted with the robot, Kismet's behavior conveys a sense that it is a living being of a tender age. Women especially speak to Kismet in "motherese," the exaggerated style they might use with a very young child. According to Breazeal, some even empathize with the creature, "often reporting feeling guilty or bad for scolding the robot and making it 'sad.'"

Equipping Kismet with appropriate facial expressions might seem like a trivial engineering project compared to the difficult one of making ASIMO walk properly, but actually it requires substantial hardware. Moreover, Kismet's actions must be integrated with the robot's perceptions to give responses that make sense to people, which requires a large dose of social or interpersonal intelligence.

The electronic brain behind that interpersonal intelligence could not fit into Kismet's head. It consists of 15 external networked computers that give Kismet "drives"—that is, built-in needs to be with people, and to be stimulated by them—and "emotions." Each drive has a preferred operating point; for instance, to be neither over- nor understimulated. Depending on how well these ideal conditions are met as Kismet interacts with someone, the robot enters into one of three arousal states—bored, interested, or calm—and its face shows anger, disgust, fear, joy, sorrow, or surprise. Humans respond to these cues by modifying their conversation and actions; for instance, speaking soothingly if Kismet is angry or overstimulated, or waving a toy if it is bored. At their most empathetic, the human responses maintain Kismet's drives at their ideal points, which puts the robot into a state of well-being; it is not overwhelmed, yet it is challenged to interact further—exactly what a good parent or teacher hopes to give a child. Breazeal has set up a complex, socially based feedback loop between robot and human that works to keep them mutually engaged.

ASIMO was built to emphasize kinesthetic intelligence, and Kismet the interpersonal type. A third smart robot that displays these two categories of intelligence and more has been developed by the Sony Corporation, maker of the smart robotic dog AIBO. Sony began working on its humanoid Sony dream robot (SDR) in 1997. The latest version, significantly advanced, was originally called the SDR 4X II but now has the catchier name QRIO (Curio). It extends the kinesthetic intelligence of ASIMO and adds interpersonal, linguistic, and even musical intelligence in a unit only 58 centimeters (23 inches) tall. At the ROBODEX 2003 show, the broad abilities of this small unit made it the natural choice as master of ceremonies for the parade of varied robots, most of them much larger, that enlivened the event.

QRIO is literally 50 percent smarter than the preceding model because its brain consists of three microprocessor chips rather than two, all of which are specially optimized for high speed. The third chip provides speech synthesis and recognition, replacing an earlier external computer. The improvement in vocabulary is dramatic, from 20 words to at least 20,000, with which the robot is able to hold simple conversations. It also displays a degree of musical intelligence: It sings as well as talks, and can do so in harmony with other QRIO units. Its social interaction abilities, combining visual cognition and linguistic capability, have also come a long way: It greets people it knows by name. It learns new people by asking them to "Please hold still for a minute" then, using its dual camera stereoscopic vision and hearing, memorizes their faces and names. Once a face is learned, the unit can pick it out from a crowd.

The robot also has excellent kinesthetic intelligence along with a superior physical endowment. As I learned from Masahiro Fuita of Sony's Digital Creatures Laboratory, one of the prime designers of the robot, specially designed actuators give it a particularly smooth gait and motion. In addition, sensors that feed it kinesthetic information keep it well balanced, even to the extent of adjusting its walking style depending on the surface under its feet. In a demonstration by Sony, one of the unit's most lifelike acts was to squat down as it moved from one kind of flooring to another, examine the new surface, and then resume walking with a different gait. It is extraordinarily agile as well, moving with enough speed and coordination to dance—another facet of musical intelligence. And it has sufficient spatial intelligence to plot a path and avoid obstacles as it walks.

QRIO and its predecessor, Sony's AIBO dog, are carefully crafted to mesh with human expectations. Like Cynthia Breazeal, the design team at the Digital Creatures Laboratory and its consultant Ronald Arkin at Georgia Tech, draw on theories of emotion, and on ethology, the study of how living species behave in their natural settings. Insights from these areas are translated into programming architectures that create believable behavior patterns. (Interestingly, the designers note that ethological information for humans is more limited than for dogs, mostly because of privacy issues.)

These pioneering researchers have constructed a psychic space between human and robot where the person naturally tends to ascribe emotions and other characteristics of living beings to the creature. In addition to behaving in a variety of intelligent ways, the three smart beings, and others, seem to have emotional components too.

THEY THINK . . . BUT DO THEY FEEL?

With their social capabilities, ASIMO, Kismet, and QRIO all show what look like emotions or frames of mind, although to vastly different extents. ASIMO's aspect is similar to a spacesuit helmet visor rather than a face, but its confident walk and jaunty hand wave suggest a certain attitude. QRIO has a face, with appealingly big eyes, but the features are immobile; still, it can simulate emotional expressions by singing, dancing, and conversing appropriately.

Kismet shows emotions directly, however, in its face and voice, and as we have seen, these expressive features can make a bond between human and robot. Because Breazeal also uses the words "emotions" and "drives" to describe Kismet's internal workings, it is natural to ask whether they have meaning to the robot itself—does it have innermost subjective feelings beyond what people ascribe to it? To put it another way, some people empathize with Kismet, feeling bad when they make it "sad." But is anything like sadness going on inside Kismet? And might such feelings hint at intrapersonal intelligence, Gardner's seventh category?

To Breazeal, the answer is unequivocally no. She writes, "Kismet is not conscious, so it does not have feelings. . . . That Kismet is not conscious (at least not yet) is [Breazeal's] philosophical position." Rodney Brooks agrees. "It is all very well," he writes,

> for a robot to *simulate* having emotions . . . it is fairly easy to accept that [roboticists] have included *models* of emotions . . . some of today's robots and toys *appear* to have emotions. However, I would think most people would say that our robots do not really have emotions. (Italics in the original.)

Brooks's viewpoint is echoed by other researchers, such as Rosalind Picard of the MIT Media Lab. Picard is a pioneer in affective computing; as she defines it, "computing that relates to, arises from, or deliberately influences emotions." She believes that "Computers do not have feelings in the way that people do . . . computers simply aren't built the way we are to have that kind of an experience."

Coming from robotics and computer experts, these comments about the lack of internal emotional states suggest that intrapersonal intelligence does not exist among artificial beings. But there are reasons to think that at least a low-level intrapersonal component is achievable, and that such an advance would represent an opening wedge for self-awareness, as we shall see later. Whether real or not, however, some familiarity with emotions can be more than a frill for artificial beings. If a being can smile at you, and recognize your own smile, your interaction is likely to go more smoothly than without these human attributes. This is also true for human–computer interactions: Picard notes that the appearance of emotions could, for instance, enhance computerized tutoring.

For artificial beings that move in the world, something akin to emotion can even be a survival factor. In his book *Robot: Mere Machine to Transcendent Mind,* Hans Moravec argues that advanced robots operating in the real world would need to deal with contingencies, and could do so through internal functions that parallel what real emotions do. These functions would take the form of "watchdog programs," constantly operating within the robot to keep an eye out for trouble. If such a program senses danger ahead,

> it may switch the robot from seeking its destination . . . to abruptly halting . . . and slowly, carefully, backing away from the hazard or . . . retreating as quickly as possible. The robot would react in a way we recognize as fear when we see it in animals.

Similarly, Moravec postulates a doglike robotic love or loyalty, meaning that the unit could discern which of its activities especially please its human master, and would modify its behavior to keep its human happy. Anger, too, would play a role, for instance, in the behavior of a robot security guard. Upon detecting a human intruder, the being would move from requests for cooperation, to threats, to aggressive action. Carried to its logical though horrifying extreme, this scenario would resemble the scene in the film *RoboCop* where a police robot escalates its demands so aggressively that it shoots an innocent bystander before he can possibly comply. That response forces the corporate executives to decide they need a cyborg cop with human rather than robotic judgment.

There are obvious dangers in giving an artificial being emotions or their simulations, but surprisingly, it might also be that emotions are absolutely essential for a creature to think intelligently. As Rosalind Picard notes in her book *Affective Computing:*

> In normal human cognition, thinking and feeling are partners. If we wish to design a device that "thinks" in the sense of mimicking a human brain, then must it also "feel?"

Picard alludes to the fact that what seems to us an ingrained and strict distinction between rational and emotional thought—with the latter often dismissed as somehow less meaningful in human cognition—seems not to represent how the brain really works.

Varied evidence, from studies of people with damage to specific areas of the brain, to data from brain scans, shows intricate connections between the cortex, the part of the brain traditionally associated with rational thought, and the limbic system of the brain, parts of which are associated with the emotions. According to Antonio Damasio, whose researches and writings have been seminal in developing this view, "feelings are a powerful influence on reason . . . the brain systems required by the former are enmeshed in those needed by the latter." High-order cognitive functions can be shaped and even enhanced by emotions. Emotional states affect what the senses perceive; more surprisingly, too little emotion can prevent someone from narrowing down a menu of alternatives until a valid and rational choice emerges.

The links between reason and emotion might upset a treasured science-fiction notion, expressed in Isaac Asimov's *I, Robot* and other stories, that an unemotional artificial brain can make better decisions for humanity than humanity, burdened by emotions, can make for itself. This perception might prove to be an illusion if machine emotions are necessary to produce fully capable machine intelligences. If that is the case, it adds a new dimension to the construction of an artificial brain: It might be necessary to ensure that the brain includes whatever combination of rational and emotional systems is necessary for full creative thought and consciousness.

Even without emotional components, however, it is a challenge to make digital microprocessors that process information as effectively as the human brain. New designs and modes of operation of computer chips might be required to improve sheer processing power, and eventually to incorporate new functions such as emotions. As Rosalind Picard has pointed out,

> There are certain ways in which emotion influences memory . . . that are not as obviously easy to implement in present machines . . . I would encourage some radical architectural rethinking and probable redesign.

No one seems to be actually working on an "emotion chip" like the unit used by the android Commander Data in the *Star Trek* series, but researchers are working to improve artificial brains by increasing the speed and capacity of present-day digital chips, and by designing new types of chips. Further off on the horizon is the possibility of using living neurons or brains in cyborglike arrangements. To understand these possibilities, we need to examine both organic and artificial brains.

GRAY MATTERS

Look at the surface of a human brain and you see gray, distinct from the innermost parts that are white. The gray surface is the cortex, a thin layer (6 millimeters or 0.25 inches thick) covering the brain ("cortex" comes from the Latin word for tree bark). It is the area where, according to traditional views of the brain, much of our thinking goes on; hence, the association of "gray matter" with intellectual activity. The cortex consists of layers of neurons, along with other cells that give them physical and physiological support. The neurons are arranged in orderly fashion, lying in vertical columns that are further grouped into functional clusters such as the visual and auditory cortices.

Much of the power of the brain comes from the sheer processing power of its 100 billion neurons. The cortex contains more of these than you might think because its wrinkles disguise its large surface area. Smoothed and spread out, the cortex would cover a square yard. Also significant is the number of interconnections among the neurons. Each neuron is connected to a thousand or more others, made possible partly by the fact that they link through three dimensions, vertically as well as horizontally.

Artificial brains—that is, microprocessor and memory chips that manipulate and store data—are also a form of gray matter, of a shinier sort, the color of pure silicon, polished to a reflective gunmetal sheen. Like neurons, the interconnected transistors etched into a chip transfer signals among themselves and come in huge numbers. Compared to a computer chip from the late 1970s, a modern Pentium processor has a thousand times more transistors and manipulates bits several thousand times faster, typically executing about 2,000 million instructions per second (MIPS). Today's memory chips typically hold hundreds of megabytes of data.

But even the millions of transistors in a computer chip hardly compare to the billions of neurons in the brain. Further, the transistors are interconnected less abundantly than are neurons, partly because the chip is flat and not three-dimensional, and partly because it works serially, by executing one instruction after another. That scheme is less powerful than the massively parallel multiple processing carried out by an organic brain. In this respect, chips are inferior to natural brains, but they are also faster, transferring electronic signals in nanoseconds, a million times faster than organic systems. This enormous edge in speed makes it possible to simulate in real time much of what the brain does, although the basic designs are different.

There is no question that artificial intelligence will grow as the processing power and storage capability of computer chips increase. But can chip-based intelligence reach the full power of the human brain? According to Hans Moravec, that goal would require microprocessors running at 100 million MIPS—50,000 times faster than a Pentium chip—supported by memory chips that store 100 million megabytes. Today's most sophisticated computers, such as the unit that currently holds the world's speed record—the Earth Simulator computer in Japan, which models our planet's global behavior—are approaching this kind of speed. Even the Earth Simulator, however, is not remotely likely to form the brain of a mobile being: It cost about $500 million and occupies an entire building. The trick is to achieve the requisite speed in a tiny, low-power microprocessor chip. While that might seem a fantastic extension of what a Pentium chip can do, the pace of improvement has been staggering, with chip speeds roughly doubling every 18 months, so there is hope that the appropriate level can be achieved with our present silicon technology, perhaps in the 2020s as Moravec predicts.

But improvements in nanoscale silicon technology might run into roadblocks set by the behavior of matter at microscopic scales, which scientists do not fully understand, or by engineering problems, such as dealing with the heat thrown off by myriad transistors. If one of these roadblocks proves impassable, other approaches to nanoelectronics currently under study might still lead to more powerful artificial brains. A single molecule can now accomplish what a transistor does, suggesting the possibility of very compact processors and enormous memory banks. Scientists are also working on a new, exceedingly powerful type of computing based on quantum mechanics that might become a practical reality.

So far, these new approaches to computation are only in the experimental stage. Researchers are looking at other designs for building brains that bypass the limits of conventional chip technology, hoping to match the effectiveness of natural brain processes. One alternate approach is to replace standard digital chips with neuromorphic chips—also made of silicon, but with a different mode of operation, one that mimics the workings of neurons and organic brains, which do not operate digitally. The transistors in a digital chip function as on-off switches that represent the binary digits 1 and 0, whereas the currents that flow among neurons do more than turn on and off. In neurons, the magnitude of the currents and how the currents change with time are essential information. Data transmitted in this way are said to be analog rather than digital in nature.

The differences between electronic digital processing and biological analog processing motivated two founders of the so-called neuromorphic approach: Carver Mead, Professor Emeritus at Caltech, who coined the term in the late 1980s, and Eric Vittoz of the Swiss Federal Institute of Technology. Compared to digital technology, writes Mead, "Biological information-processing systems operate on completely different principles," and adds:

> For many problems . . . biological solutions are many orders of magnitude more effective than those . . . using digital methods. This advantage can be attributed principally to . . . the representation of information by the relative values of analog signals . . .

For Vittoz, an important advantage is that, "The collective computation carried out by the brain in its massively parallel architecture can be emulated on silicon."

Kwabena Boahen, at the University of Pennsylvania, uses neuromorphic principles in the synthetic retina mentioned in the previous chapter, and emphasizes the efficiency of the method for the construction of an artificial brain. Transistors in a digital mode operate faster than neurons, but they also use more power. Boahen estimates that a digital chip as capable as the human brain would consume a billion watts of electrical power. If his estimate is anywhere near accurate, it is highly unlikely that we could ever build a humanlike digital brain in a form suitable for an artificial being. To explore the possibilities of using neuromorphic chips that copy how the brain operates, researchers in Boahen's group have projects under way to develop a chip that follows the special pattern of connectivity among the six layers of neurons in the visual cortex, and to implement the thousandfold connectivity among neurons that characterizes the brain.

Another new design with similarities to the brain is the adaptive or reconfigurable chip. In standard chip technology, once the transistors and other electronic elements are etched into the silicon, the circuitry they form is physically fixed, which also fixes what the chip can do. Normally the only way to change the circuit is to make a new chip. An adaptive chip, however, contains huge numbers of fundamental data-processing elements. Like myriad Lego blocks connected to form any one of an infinite number of structures, these elements can be connected in different configurations to perform different functions. Remarkably, this can be done on the fly with software commands rather than by the laborious process of changing hardware. Paul Master of QuickSilver Technology, a firm that along with Intel, IBM, and others is investigating adaptive chips, says: "Until now, the hardware had to match the problem. Now we can change that."

According to proponents of the technology, adaptive chips offer speed and efficient use of power, in themselves important factors for artificial brains. But the truly novel aspect of the adaptive chip is its ability to change itself in real time, a crucial feature of the human brain not available in conventional chips: plasticity, the ability to change neural pathways according to new experience. Plasticity plays a large role in our development when we are young, and in learning and memory throughout our lives. It also gives the brain a way to compensate for functions that have been lost or reduced due to injury.

At this point in the development of adaptive chips, potential applications are limited to specialized devices such as cellphones, and there are questions about the long-term impact of the technology. But if it proves viable, then as is usually the case in nanoelectronics, improvement will come at a furious pace. We might expect to see second-generation silicon brains with the built-in ability to change and learn just as the human brain does—or perhaps solve for themselves the essential problems, such as finding the right mixture of rational and emotional thought.

There is, however, a different direction to consider in making brains for artificial bodies, an ultimate possibility that avoids the difficult task of re-creating in silicon what nature has already worked out. This is the idea of merging living neurons or brains with artificial systems to create bionic or cyborglike arrangements. Whatever the advantages of neuronal processing, whatever the correct mix of reason and emotion, whatever the combination of brain structures and interactions that gives rise to consciousness, they already exist in living neural systems and brains.

The connection between organic brains and artificial constructions requires brain–machine interfaces (BMI), a technology still in its infancy. If true cyborgs, neural prosthesis, or the "jacked-in" world described by the science-fiction writer William Gibson ever come to pass, they will start with this research.

CYBORG SCIENCE

It is a serious step, scientifically and ethically, to go beyond emulating or copying the brain in silicon to merge living nervous systems with electronic networks, or to use living brains to operate nonliving devices. Some people feel a visceral shudder at the idea; it seems an unholy alliance, creating for them the same dread Victor Frankenstein felt as he animated a dead hulk. Grisly though the merging might seem, it is hard to argue with the main motivations for exploring this possibility; namely, the desire to aid paralyzed and handicapped people and also to develop new probes of the brain. Most of the scientists and physicians who work in this field (and the number has grown explosively in the last decade) seek to improve life for the ill and handicapped, but their research potentially also forms the basis for cyborg science.

One approach to making brain–machine interfaces focuses on the brain-wave method, where electrodes pressed against the scalp detect the brain's neural activity, and the signals detected in this way are connected to an appropriate interface and computer. The information gained thereby can be used to train patients, who learn to tailor their brain activity to perform such useful functions as guiding a computer cursor. The brain-wave method requires no brain surgery; however, it receives information from many neurons, a mixture that is difficult to interpret and put to optimum use. The pure signal from a single neuron is easier to analyze. For this reason, many BMI researchers want to implant electrodes in the brain to obtain well-defined signals from individual neurons. Several of these efforts are distinctly cyborglike, using a living brain to control an external device.

A case reported in 2000 might be the closest we have yet come to a natural brain devoted to an artificial body, like the dancing cyborg Deirdre in C.L. Moore's story "No Woman Born"—except that this real-life cyborg did not use a *human* brain. Sandro Mussa-Ivaldi and his team at Northwestern University Medical School wanted to learn about connections between sensory information and motor responses in organic brains, so they created an animal cyborg. Its brain came from a sea lamprey, a predatory eel-like fish, chosen because its neurons are large and easy to manipulate. The portion used for the cyborg was the brainstem, which deals with vision and balance and ordinarily issues motor commands to control the animal's swimming movements—but in the experiment, the fish's body was replaced by a small two-wheeled mobile robot equipped with a set of light sensors. The brain, kept alive in a nutrient solution for experiments that ran as long as 8 hours, was connected to the light detectors and to the motors controlling the wheels by means of implanted electrodes.

To test the contrivance's brain–body interaction, it was placed in a small arena ringed by light sources. As the lights were switched on, the brain received signals from its light sensors and responded by sending signals to the wheel motors, causing the robot to move. Not every installation of electrodes was successful, but when an installation worked well, the device moved consistently either toward or away from the lights, depending on how the electrodes were placed. In short, the artificial body was under the control of an organic brain that produced meaningful motor responses to sensory information, acting in a limited but truly cyborgian way.

Remarkably, cyborglike activity has also been achieved with the more powerful brains of monkeys and humans. In 2000, a group led by neuroscientist Miguel Nicolelis at Duke University used the brain of a living monkey to control a robotic arm. The researchers implanted an array of fine wire electrodes in the brains of two owl monkeys. First they used the array to record neural patterns as they trained the monkeys to use their hands for specific tasks, such as reaching for food. These patterns gave enough information to predict where a monkey's arm would be, milliseconds after the monkey's brain produced the signal. The scientists used the data derived in this way to write a control program for robotic arms physically separate from the monkey. One arm, located in the same laboratory, was directly connected to the brain: a second one, distantly located in a laboratory at MIT, was connected via the Internet. For each arm, as the monkey moved its natural arm, the artificial limb faithfully followed its movements.

This astonishing result is not the final answer to the problem of giving a paralyzed person a useful BMI, which ideally should tap neural signals without any need for bodily motion. In 2003, Nicolelis and his group reported achieving that goal, although in monkeys, not humans. The scientists implanted electrodes in the brains of two macaque monkeys and recorded the neural patterns that arose as the animals learned to work a joystick that controlled a computer cursor; the monkeys were rewarded with a drink of juice when they used the cursor to reach and virtually grasp a target on the computer screen. Then the researchers disconnected the joystick but continued to tap the neural signals. Baffled at first, the monkeys soon realized that although manipulating the joystick no longer did anything, they could guide the cursor merely by thinking about it, with no muscular action. In a final step, the signals from the brain were rerouted to control a robot arm rather than the cursor. "By the end of training," says Nicolelis, "I would say that these monkeys sensed they were reaching and grasping with their own arms instead of the robot arm."

The ultimate goal is to develop similar brain–machine interfaces for humans. As an important step toward that goal, the Atlanta-based researcher and neurologist Philip Kennedy has, for the first time, enabled a human brain to control an external device with no associated bodily action. Kennedy implanted electrodes into the brain of a paralyzed man, who then learned to control a computer without moving a muscle.

Much of Kennedy's work began at Emory University, and continues at his company Neural Signals. A visit to the company mostly reveals cluttered electronic workbenches where Kennedy and his colleagues work out how to process the signals elicited from implanted electrodes. But the heart of Kennedy's achievement is the unique "neurotrophic" electrode that he developed, so small that it is not apparent on his workbench. Its design keeps it well anchored in the brain without undesirable motion or loss of electrical contact, problems that can plague conventional electrodes. The device consists of two fine gold wires attached to a tiny hollow glass cone the size of a pen tip. The inside of the cone is coated with nerve tissue, taken from another part of the patient's body, which encourages the growth of nerve cells. When the cone is inserted into the brain, neurons grow through both its open ends to hold it firmly in place as they connect to the electrodes. The signals from the wires are sent first to an amplifier and then to a computer.

Kennedy's great achievement has a second part, in clinical settings. He has implanted his electrode in patients who were severely paralyzed due to various causes including amyotrophic lateral sclerosis (ALS or Lou Gehrig's disease), the condition that afflicts the British physicist Stephen Hawking. Among these patients was Johnny Ray, a Vietnam War veteran who was left "locked-in" at age 53 by a stroke; that is, he was left with full cognitive function but no bodily control except for limited face and neck motion. In 1998, Kennedy implanted two electrodes into Ray's motor cortex (the part of the brain where movement commands originate as its neurons fire), specifically in the area devoted to moving the hands.

Ray sent signals through the electrodes to control a cursor he could watch on a computer screen, but achieving that control was not an overnight process. In initial training, Ray was asked to imagine he was moving his hands to activate the electrodes, but this strategy did not produce consistent results. After some months, however, he was able to move the cursor reliably, and these movements were correlated with movements of his face and neck. Finally, Ray was able to accurately place the cursor wherever he wanted to with no bodily motion at all, which showed that he was controlling the cursor with his brain. When Ray reached this point, Kennedy asked him what he thought about as he moved the cursor. Ray's reply (spelled out letter by letter with the cursor) was that he thought only about moving the cursor and not about any part of his body; so from Ray's own viewpoint, he moved the cursor by mind alone.

One of the more interesting results of this feat relates to how the electrodes interact with the plasticity of the brain. Kennedy interprets his results as showing that as Ray was trained, the implanted "hand cortex" changed to "cursor cortex," devoted only to cursor movement. Based on his work with monkeys, Nicolelis makes a similar point:

> Long-term operation of [a neurally controlled prosthesis] by paralyzed subjects would . . . through a process of cortical plasticity . . . confer to subjects the perception that such apparatus has become an integral part of their own bodies.

If human neurons actually adapt themselves in this way, that offers another potential benefit to patients; it gives great freedom as to where electrodes can be placed in the brain, regardless of where the brain is damaged. Changes in the brain due to neural interfacing might have other implications, too, that warrant careful consideration.

Although implanted electrodes enable a brain to communicate with the exterior world, efforts also continue to forge BMI connections that do not need surgery. One such BMI technology being pursued under DARPA auspices is based on magnetic effects. According to DARPA, the aim is to "communicate with the world directly through brain integration with peripheral devices and systems." For example, pilots could control aircraft just by thinking about how they want them to move (as in the 1977 novel, *Firefox*, by Craig Thomas, and the 1982 film of the same name) or an infantryman could operate a powered exoskeleton. It has been known for some time that neural activity produces magnetic fields that extend through the skull and can be measured by sensitive detectors. Many questions remain, though, such as whether the communication can work in both directions; that is, both from and to the brain, and whether the technique can be made fine-scaled enough to pick up signals from individual neurons. But the achievement of a noninvasive BMI would give a different meaning to bionic enhancements for civilians as well as soldiers; instead of submitting to a serious surgical procedure, a person could freely add or remove neural prosthetics, such as auxiliary memory or communication devices.

Other researchers are working on truly radical methods to interface neurons with electronic devices. One pioneer, Peter Fromherz of the Max Planck Institute for Biochemistry in Martinsried, Germany, aims to connect individual nerve cells with transistors by creating a "neuron on a chip." He sees possibilities for neural prostheses and BMI units, and for new ways to study the brain, but there are difficulties. Although both chips and neural structures encode information electrically, the mechanisms differ. A chip carries electricity by means of a lightning-fast flow of electrons in a solid, whereas a neural system carries electricity by sluggish electrochemical means involving the movement of ions—charged atoms—in fluids. Hence the two systems operate at different time scales, nanoseconds for chips versus milliseconds for neurons. Finding ways to interface the systems is one challenge; ensuring that the silicon provides a proper base for nerve cells to grow and adhere is another.

Despite these challenges, Fromherz has created what might be called nanocyborgs, that is, prototypical neuron–chip hybrids, which consist of neurons from rat brains, leeches, or snails specially grown atop silicon chips so that the neurons overlay the transistors etched into the chips. Fromherz's initial experiments show that the natural and the artificial systems indeed communicate directly; that is, electrical activity in the neuron causes activity in the chip, and vice versa. In a more complex demonstration, when two snail neurons were placed some distance apart on a silicon chip, the firing of one neuron induced the other to fire, although they were not linked by neuronal means but only through the silicon. Fromherz has also extended his technique to entire sections of a brain. In a structure consisting of a thin slice of rat brain on a silicon chip containing a row of transistors, when the brain was electrically stimulated, so was the entire array of transistors.

These early, though encouraging, results are providing insight into the linkage between cell and silicon. Fromherz calls the two-neuron experiment "a silicon prosthesis," suggesting that it might some day be possible to repair neuronal circuits electronically. Nevertheless, he urges caution, writing that "visionary dreams of bioelectronic neurocomputers and microelectronic neuroprostheses are unavoidable and exciting, but they should not obscure the numerous practical problems."

SIGNS OF CONSCIOUSNESS

A neuron on a chip is not conscious—although a network of them might be—but the other hybrid beings we have considered are. A locked-in paralyzed person like Johnny Ray brings a mind with subjective experience, a sense of self, and a personal history to whatever artificial extension is added through a BMI. It might never be desirable, ethical, or even possible to have a human brain operate a complete artificial body, rather than just a computer cursor or a mechanical arm. But if that were to happen, the result would be a being with the conscious personhood of the original mind (in Howard Gardner's terms, intrapersonal intelligence), and the physical attributes of the new body—just as the cyborg dancer Deirdre kept her essential personality even as her brain functioned within a metal shell. (However, according to the findings that the brain changes as it operates new devices, that personality might change as well, as was feared would happen with Deirdre.)

Next, consider a hybrid being many steps below human mental capacity, a lamprey brain operating a robotic body. Does that cyborg possess consciousness? Scientists and philosophers would agree that even a living lamprey with brain and body intact lacks a sense of personal existence, what the consciousness theorist Gerald Edelman calls higher-order consciousness. But it does possess what Edelman calls primary consciousness, the lower-level, nonlinguistic awareness of its immediate world through sensory information. The chunk of lamprey brain that operates the robotic body is even less of an individual being; nevertheless, in responding to sensory input and reacting by moving its artificial body, it shows primary awareness.

Now remove any living element and consider a completely artificial being. Does it possess consciousness? For the three smart robots we have examined, ASIMO, QRIO, and Kismet, there is no evidence for subjective experience, which is essential for higher consciousness and is implied in Gardner's intrapersonal intelligence. But although the digital creatures are unaware of a self, they show low-level consciousness like that manifested by the lamprey cyborg. Through their senses, they know the world and respond to it.

They also have something else, internal body awareness through kinesthetic intelligence. As Hans Moravec points out, that kind of knowledge could easily be enhanced by designing a unit that registers, for instance, the power level of its own batteries and the operating temperature of its motors. Physical self-knowledge combined with knowledge of the world does not constitute consciousness of self, but if the unit uses information about its internal states to plan its actions, then it is displaying a salient characteristic of higher consciousness—namely, projection into the future—rather than responding only to the present moment. This state of consciousness is a more elevated one than that evinced by the lamprey cyborg.

There is another way that self-knowledge begins to touch on higher consciousness in artificial beings. According to theories of emotion, an event that elicits emotion triggers activity in the autonomic nervous system—the involuntary part of our nerve network that controls the glands, the heart, and more—to physiologically prepare the body to adapt to the stimulus. A person might experience nervousness by sensing the physiological reaction of "butterflies in the stomach," along with a sense of fear. This might seem irrelevant to a robot like Kismet, which lacks any physiology. However, as Cynthia Breazeal points out, Kismet's complex programming includes something roughly equivalent—a quantity that specifies its level of arousal, depending on the stimulus it has been receiving. If Kismet itself reads this arousal tag, the robot not only is aroused, it *knows* it is aroused, and it can use this information to plan its future behavior. This, too, begins to resemble an important characteristic of higher consciousness; namely, the reflexive ability of a mind to examine itself over its own shoulder.

Unquestionably, these examples are a far cry from Gardner's broad vision of intrapersonal intelligence, or that sense of selfhood projected forward and backward in time that defines the higher consciousness. Yet these tiny first steps toward self-knowledge might be the beginnings of an evolution toward full digital thought and consciousness—an evolution that has only begun, and whose prospects, along with other facets of the future and meaning of artificial beings, we examine in the last chapter.

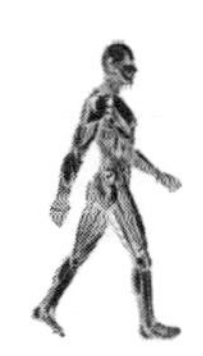

9

Frankenstein's Creature or Commander Data?

The development of advanced artificial beings and of bionic humans is well under way. The pioneering efforts of roboticists, neuroscientists, and other researchers are creating a powerful cross-disciplinary technology for the coming century, a rich medical–technical environment that might lead to autonomous artificial beings and to enhanced human bodies and minds. This technology is actively driven by a variety of motives: scientific curiosity and the technological imperative, benefits for human health and longevity, and applications in areas from industry to space exploration to warfare.

At the moment, industrial robots dominate: The latest comprehensive survey *World Robotics 2002,* issued by the United Nations Economic Commission for Europe, puts their world population at 760,000, projected to soon reach a million. That same report, however, also predicts increased use of robots in areas such as medicine and security, and explosive growth for household and entertainment robots, with a hundredfold increase in units sold between 1999 and 2005.

Despite this growing activity, no one has yet made a completely autonomous being, or one that seems consistently and convincingly alive, or a bionic implant that improves human strength or wit, or a true cyborg, a living brain in a mobile artificial body. But there is no doubt that existing technology will carry us further along these paths. At the physical level, the creation of walking robots has taught us a great deal about mechatronics and body construction. Devices for implementing artificial senses, from light and sound detectors to wireless receivers, are also well developed and will only get better. Many issues about the physical capabilities of artificial beings—notably, how to extend their battery-powered lifetimes so that they don't need frequent recharging—remain, but we do have clear directions for body improvement that apply known principles without having to invent new concepts.

Neither artificial bodies nor synthetic senses can work meaningfully without guidance from a brain, a mind, or a developed cognition. Here, too, progress will come through the refinement and evolution of the existing approach, which is to program digital computer chips to simulate what the brain does. Every increase in hardware speed and capacity, and in the cleverness of the software, makes artificial beings more effective, just as the addition of a third chip to Sony's QRIO robot enormously enhanced its speech. But a deeper understanding of our own brains, leading to the construction of better synthetic ones, might be needed to bring those silicon brains to a new plane—truly high intelligence, and possibly silicon-based consciousness.

Human–machine connections have bright prospects as well. The potential medical benefits are clear. We will see rapid progress in this area, from improved cochlear implants for the deaf to more effective visual replacements for the blind and better BMI technology for the paralyzed, perhaps leading soon to direct neural control of an artificial limb. These achievements form a basis for the next level, which would go beyond healing to extend human mental and physical abilities—for instance, by connecting a human brain directly to the Internet or to a powerful computer, or permitting the brain to directly control a complex device such as a weapon or an artificial body. Because much of the current research in these areas is funded by the Department of

Defense, it is possible that scientists have already made dramatic progress that is being kept secret, but as far as the open literature shows, we are not close to achieving these science-fictional possibilities. However, serious research along these lines is just beginning.

Neither the building of artificial beings nor the creation of hybrid humans is just a matter of getting the technology right. Even if supernatural fear of synthetic life is long gone from our psyches, we are still concerned about what this technology means for people, and we need to answer some questions that have profound implications: What is our purpose in making artificial or hybrid beings? What are our ethical responsibilities toward them, and theirs toward us? Do we have anything to fear from intelligent and powerful nonhuman beings—if not the violent overthrow of humanity portrayed in Capek's *R.U.R.*, then more subtle damage such as debasing human worth or causing economic harm? Is a hybrid being, part human but perhaps mostly machine, still a person, or something else, and can a fully artificial construction be a person? If we learn to enhance human health or mental ability by implantation, who should receive these benefits?

These questions have different answers for different societies; for instance, in Japan, where robots are developed primarily for civilian use, and the United States, where military applications of robotics play a large role, and so the answers, like the beings themselves, reflect light back onto our own nature. Many of these issues will not arise, however, until artificial beings become more capable than they are now, and that means becoming more intelligent.

GETTING SMARTER

Some researchers are confident that digital chips will eventually attain the full power of the human brain, at least as judged by a quantitative measure—making the chips operate so fast that they match the speed of the brain's extraordinary parallel processing. As we have seen, Hans Moravec estimates that a microprocessor running at 100 million MIPS would be as capable as the brain. In *The Age of Spiritual Machines,* the inventor and computer visionary Ray Kurzweil considers the same question. He differs from Moravec in his estimate of just how much processing the human brain does, but both men predict that the present rate of chip development will take us there in 20 years or less. Kurzweil puts it dramatically, predicting that by the year 2019, a mere $1,000 will be enough to buy the computing power of the human brain.

Even if this quantitative success is achieved, will it produce a brain that can sustain the full equivalent of a human mind and consciousness? To many builders of artificial beings, this is not a key question. The immediate goal of those researchers is the construction of beings that behave in ways that are or appear to be intelligent, emotional, social, and whatever else is useful, without insisting that the beings think like humans or worrying about "real" emotions within their silicon brains. Faster computation will accomplish that much, if not through evolutionary improvement, then through advances such as molecular-level or quantum computing.

No matter how rapid the computation, beings based on computer-style processing might end up thinking like...well, computers. This is not to say they won't be effective; in fact, they might well surpass humans in many ways. If there is to be a next stage, however, where an artificial being acts with full autonomy, shows intrapersonal intelligence, or looks you in the eye and announces "I'm conscious," we might need to consider qualitatively different methods of constructing artificial brains.

Still, the first step toward more capable beings is to extend the artificial brains we already have, which are based on programmed digital chips. These brains are showing signs of Howard Gardner's multiple intelligences, but except for logical-mathematical intelligence, the artificial beings controlled by these brains operate mostly at the level of a young child and have yet to achieve a meaningful degree of intrapersonal intelligence. Nevertheless, there are steps we can take to make these beings more capable and more complete.

One step is the improvement of their bodily-kinesthetic intelligence. It took considerable time and work for that kind of intelligence to reach its first major success with the construction of the walking ASIMO robot and others. The next kinesthetic goal should be further development of autonomous grasping and manipulation. Some robots perform these tasks at a simple level, but only with pincerlike hands. And although some units, such as NASA's Robonaut, have dexterous fingers-and-thumb hands, they require a human operator. Complete hand intelligence would be an important step toward more useful beings; also, manual dexterity would give these embodied intelligences a way to explore and shape the world, developing their brains in the process. Such an extension of bodily-kinesthetic ability requires better spatial intelligence and has to incorporate object recognition, another ability that falls under spatial intelligence and is now under development.

Artificial musical intelligence is already here, and not only in the singing and dancing that the Sony QRIO robot performs—some computers are already composing music. Why not add this ability to QRIO? Of the remaining three categories of intelligence, the interpersonal type will also develop as artificial systems become better at distinguishing human emotions as expressed in the face and voice, and responding in humanlike ways. But linguistic intelligence and intrapersonal intelligence—or self awareness—raise special issues.

BAD LANGUAGE

Linguistic intelligence is exceptionally significant in evaluating artificial beings because of its role in the Turing test. The ability to use language might lie at the pinnacle of human intellectual functions; in fact, some theorists hold it to be essential for our very thoughts. Thinking and self-awareness can be seen as a process of narration and response that we carry on inside our minds, a dialogue in an internal voice that is the core of the "I" within each of us.

The Turing test recognizes the importance of language, and so did those pioneers of AI who in the 1960s and 1970s tried to emulate important parts of human intelligence on computers. Linguistic intelligence entered in their attempts to make computers communicate in natural human language rather than programming language, and to translate from one human language to another. It quickly become apparent that these are extremely knotty problems, largely because the meaning of words in human language is often ambiguous and depends on context. The literature is full of amusing misreadings by machines. In his book *Mind Matters: Exploring the World of Artificial Intelligence,* James Hogan tells how in one project in the 1960s, the metaphorical phrase "Time flies like an arrow," perfectly clear to you and me, was sadly misunderstood by a computer; one of its interpretations, for instance, was "Time flies—a kind of fly—are fond of an arrow."

Efforts to enable computers to be programmed in natural language and to translate human languages continue, although they are not yet perfect. With large speech databases and fast processors, machine conversation using word recognition and synthesis is becoming routine in such applications as travel booking. What works over a telephone also works in a mobile unit, and so the Sony QRIO robot has language capability. But these systems can hold only limited conversations, a far cry from the generalized and diverse humanlike response the Turing test is meant to uncover.

In 1950, Alan Turing predicted that a computer would pass his test by the end of the twentieth century, but we are still far from developing a synthetic intelligence that can persuade us of its own personhood. The best-known attempt to determine how close we are to this goal is the yearly competition sponsored by Hugh Loebner, a hardware manufacturer who developed an interest in the problem and offers a substantial prize for the computer program that best meets the Turing criterion. Rather than using voice communication, these Loebner Prize events test linguistic intelligence by using keyboards for the human-computer interaction, as Turing envisioned.

The Loebner competitions began in 1989, and initially—especially in 1991—attracted luminaries of the AI world. But the conversational ability of the artificial minds was disappointingly poor in those first years and has not improved much since; Loebner calls the level of performance "gruesome." Some of the AI community has repudiated the competition, protesting that it is conducted in a way that renders it scientifically useless. Some also reject the validity of the Turing test for judging intelligence at all. But the test clearly has meaning, and enormous historical, intuitive, and emotional appeal. It is hard to avoid the conclusion that if the experts could have created a worthy conversational partner, they would have done so, and happily announced it—if not through the Loebner event, then in some other venue.

So the linguistic intelligence required to pass the Turing test remains elusive. Ray Kurzweil thinks the test will be passed by the year 2029, perhaps by one of his $1,000 equivalents to the human brain, but gives few specifics. Two avenues, however, are natural to pursue. One possibility depends on the fact that verbal communication carries more information than written forms: it is the idea of reducing the ambiguity of human speech by prosodic analysis, which—as we have seen—is already under development.

The second possibility, which has implications for artificial intelligence in general, not only its linguistic component, is to enormously expand the databases an artificial being needs for intelligent conversation. One necessary database is the speech corpus, which determines how many words the being recognizes and can say; the other is a database of general knowledge, essential to converse with humanlike diversity. Both can now be established at huge sizes, terabyte upon terabyte, without storing them within every artificial being, because they could be accessed from the Internet by any being with a high-speed wireless connection.

According to some researchers, a database of general knowledge is an absolute prerequisite for artificial intelligence in its broadest sense. As Roger Schank and Lawrence Birnbaum of Northwestern University have put it,

> The truth is that size is at the core of human intelligence. . . . In order to get machines to be intelligent they must be able to access and modify a tremendously large knowledge base. There is no intelligence without real, and changeable, knowledge.

Establishing sufficiently large databases, however, is still only the beginning: We do not yet know how to make a synthetic being hear any human comment and find among its databases a response that is relevant and perhaps also even passionate or humorous; or more challenging, make the being capable of initiating and leading a conversation as well as responding to what a human says.

SELF-AWARENESS REVISITED

The closer digital beings come to passing the Turing test, the better they will communicate with us, and if language is truly central to thinking, the linguistic ability that satisfies the Turing test might also be necessary for their own self-awareness. But whether or not that inner voice is essential, the human brain remains our only model for the seat of self-awareness, and its most striking feature is its complex interconnectivity. That is shown at the physical level by the convoluted structure of the brain, which reflects stages in its evolutionary history; at the neuronal level by the multitude of connections between a given nerve cell and others; and at the operational level by the elaborate network of connections and shared functions among subsystems such as the cortex and the limbic system.

This intricate arrangement is distinctly different from the linear pipeline by which computers manipulate data, suggesting that in addition to *simulating* the brain by programming digital chips, we might need to *emulate* it by using appropriate hardware, but we cannot emulate what we do not fully understand. What Marvin Minsky wrote nearly two decades ago in *Society of Mind* still applies:

> Most people still believe that no machine could ever be conscious, or feel ambition, jealousy, humor, or have any other mental life-experience. To be sure, we are still far from being able to create machines that do all the things people do. But this only means that we need better theories about how thinking works.

Because of new techniques such as brain scanning, we know more about the mind than we did then; even so, the unexplored territory is enormous. We strongly suspect, however, that the intricacy of the brain's internal interactions defines the very fabric of thought and self-perception. As Minsky puts it: "a human intellect depends upon the connections in a tangled web—which simply wouldn't work at all if it were neatly straightened out."

Gerald Edelman's theory and several others ascribe consciousness and the power of thought to those complex interactions among the brain's substructures. This points to the true importance of giving an artificial brain internal interactions such as arise between rational and emotional subsystems; namely, to copy the "tangled web" that seems to make human thought what it is. If this can be done, the next generation of artificial minds might surpass computer-bound thinking by using different types of electronic brains—be they systems that carry out parallel processing or neuromorphic chips, adaptive chips, or other architectures that follow the brain's peculiar nature.

GROWING UP DIGITAL

In addition to new designs for electronic brains, we might need something more—namely, a new philosophy—to create fully capable synthetic beings. At present, an artificial brain consists of processor and memory chips whose capabilities are firmly defined and fixed, and software—also fixed—that guides the hardware. But this is not how the human mind works. Although a newborn baby has limited abilities it has one crucial set of capacities: It can observe, interact, remember, and learn about the world. These efforts change the baby's brain through the plasticity of its neurons, and over time, the child matures into full intelligence and personhood. The child's interaction with adults plays a large role in this because it encourages adults to react to the child and teach it, and the social contact itself is necessary for selfhood to develop. Physical interaction is equally important. Just watch a tot experiment with reality as it learns to walk or carries out the experiment of throwing food onto the floor.

Brain plasticity could be emulated by appropriate hardware, such as adaptive chips. But knowledge of the world must be fed into those chips, and so artificial beings might actually need to grow into full consciousness and personhood—or to put it another way, to develop their varied intelligences—by engaging reality and socializing with people. From the psychological viewpoint, the social part of the interaction is essential. Howard Gardner writes that "highly intelligent computer programs" already exist, but considering the question of whether computers can develop personal intelligences, he comments:

> I feel that this is a category error: One cannot have conceptions of persons in the absence of membership in a community with certain values, and it seems an undue stretch to attribute such a status to computers. However, in the future, both humans and computers may chuckle at my shortsightedness.

James Hogan makes a similar point in *Mind Matters.* The difference between a human telling him "I feel the same things you do," and a machine making the same statement, is that,

> When I'm talking to a human, who I know is made like me, grew up like me, and has the same kind of accumulated cultural experience as me, I have little hesitation in accepting that the person probably feels things very much they [sic] way I do. I'm less easily persuaded when none of these things apply.

Gardner's and Hogan's remarks suggest that the best hope for the realization of truly intelligent, self-aware beings is to design them not to operate at full mental capacity the instant the power is turned on, but rather to learn as they interact with the world. Cynthia Breazeal's Kismet is an early example of a robot that deliberately follows the model of a child growing with the aid of encouraging adults. Physical interaction is equally important, to explore the world and learn from it. This is why Rodney Brooks thinks that an embodied artificial intelligence—that is, a synthetic brain controlling a body that deals with physical reality—can develop higher mental functions, an idea that he continues to investigate with the Cog robot.

The idea of an artificial being growing fully into itself is no recent invention. Alan Turing espoused this approach in his seminal 1950 paper "Computing Machinery and Intelligence," where he wrote, "Instead of trying to produce a programme to simulate the adult mind, why not rather try to produce one which simulates the child's? If this were then subjected to an appropriate course of education one would obtain the adult brain," and goes on to propose how that education should proceed.

There are older antecedents as well. Frankenstein's Being, you might recall, keenly felt his lack of nurturing and tells Victor: "No father had watched my infant days, no mother had blessed me with smiles and caresses." The Being displays a lack of social education, whereas the android Yod in Marge Piercy's *He, She and It* shows the value of this kind of interaction; Yod's connections with its maker and others give it cultural knowledge and heightened intelligence, and diminish its violent tendencies. In the film *2001* the computer Hal alludes in its "dying" speech to having been taught like a young child, although the idea is not otherwise developed. (However, although the film *A. I.: Artificial Intelligence* features a child android, it does not change in the course of the story, except perhaps for developing the desire to become a real boy.)

Mathematician Alan Turing, fantasy writers, and modern robotics engineers all come to a fascinating convergence here, illustrating the power of imaginative interdisciplinary thinking in the science of artificial beings. But important questions, not addressed in fantasy, remain: If a digital being can be made fully conscious only by having humans guide it as it grows up, what is the incentive to make such a creature? Could there be any value in investing time and effort for what might be a long, drawn-out process, which Turing estimated could take as much work as raising a real child?

If the artificial being starts with a newborn human child's ability to learn, but can do so at a far faster pace, bringing up the digital baby might be a matter of weeks, not years. Marvin Minsky has put it this way: "Once we get a machine that has some of the abilities that a baby has, it may not take long to fill it up with superhuman amounts of knowledge and skill." Still, individual mentoring seems unfeasible and uneconomic for workaday robots meant only to help around the house. The main justification for the effort would be to do everything possible to develop a truly intelligent, self-aware being, including designing a brain that knows how to learn, and committing the time and resources to giving that brain a good education.

Our future might then see two types of beings. Type I will be what we are already making, only better, with a more considerable intelligence and broader abilities, meant to assist humanity and lacking any trace of volitional behavior or consciousness. These will be true robots within Karel Capek's use of the word in the play *R.U.R.* to describe beings that are manufactured in order to work. Type II beings will exist at a higher level, designed to grow into creatures with full autonomous consciousness, using special brain hardware and human nurturance. We might ask, "Isn't this just a hard way to raise a human being?" The answer is no, and Yod the android illustrates why. Although it became more human, elements of its initial design remained. The result was a mixture of programming and free will, a blend of machine and human. This hybrid points to an exciting possibility, appreciated by creative researchers and writers alike: that silicon nature can combine with human nurture to create a unique but companionate species—intelligent, self-aware, humanlike in some respects and able to communicate with us, but with new thoughts and attitudes to share with humanity.

Imagine now a world in which we have the two types of artificial beings: those that only act as if they are conscious, and those that are conscious. We accept the fact of consciousness for the latter group, because if they have been brought up in human society, when one of them says "I'm conscious," we believe it. This is different from the Type I's, which, even if humanoid, are machines, no different from an automobile or screwdriver, and with just as little need for us to have moral concerns toward them.

Type II's, however, represent something else: a conscious spark within a synthetic body, to which we might respond by treating them like people. If this seems doubtful, consider a scenario where artificial parts can be routinely integrated into a person—let's say, to replace a gangrenous leg with a plastic one that operates under direct neural control. After the operation, the resulting bionic human is, of course, still a person in every sense. That would be true even for people who have had major physical changes such as the replacement of three or four limbs, or the kidneys, or heart, or all those. But what if a person's injured brain is repaired with a silicon prosthetic, or his entire brain is transferred into an artificial body? Is that being still a person, although perhaps a different one from who he was before?

From the physician's viewpoint, the answer is utterly clear. Philip

Kennedy, the inventor of the neurotrophic electrode, says his experience with patients like Johnny Ray has made him ask "What does it mean to be human? What does it really mean?" His answer is "As long as you've got your brain and your personality and can think . . . it doesn't matter what machinery it takes to keep you alive."

It would be no different for an internal life based in a silicon brain and existing in a body of metal and plastic—or would it? Is there a distinction between a human who has become more artificial, and an artificial being that has become more alive as consciousness is instilled? How shall we integrate beings with varying degrees of artificiality into our world, and what is our moral obligation toward them? And even for Type I robots that lack volition and free will, there remains an issue with moral overtones: For what purposes are we making them?

WE ARE THEM, THEY ARE US

Among the requirements for free will, which most of us think we have, is the ability to make moral choices. If an artificial being were to show moral judgment, that would be a strong indicator of a consciousness that humans could recognize. So far, this ability has been shown only by imagined artificial beings. When Yod the android in Marge Piercy's *He, She and It* faced the predicament of being a "conscious weapon [that] doesn't want to be a tool of destruction," it decided to destroy its maker Avram along with itself to prevent future androids being tormented by the same conundrum—just as its human lover Shira made a moral choice when she later destroyed Yod's plans. In *Star Trek: The Next Generation,* Commander Data was also capable of serious moral choices, including the decision to kill a human.

Until we have made equally sophisticated beings, however, it will remain the case that morality is not something that digital creatures bring with them, but something we give to them—through their software or hardware, as in the Three Laws imprinted in the brains of Isaac Asimov's robots or, more subtly, through our perceptions of them as good or bad, and the uses we make of them. These perceptions are not universal; they differ within different cultures. For instance, anyone who attended, like I did, the huge ROBODEX 2003 trade show and public exposition in Yokohama would have seen no reason why artificial creatures could ever be considered evil, or represent attempts to usurp God's place. In display after display from corporations, governments, and research institutions, the beings were uniformly presented as helpful to people, providing services from nursing care to home protection, or were shown as amusing and entertaining, as in a soccer game played by Sony's AIBO dogs and a quiz show featuring Honda's ASIMO. The ASIMO quiz show was played on stage with children, and though those children were actors, it was easy to see on the faces of many families visiting ROBODEX that their children thought they had entered Disneyland, only better.

Like culture heroes such as the good robot Astro Boy, this event showcased the particularly benign Japanese attitude toward artificial beings combined with Japan's leading position in robotics, which began when manufacturing robots took hold there in the late 1960s. According to Frederik Schodt's book *Inside the Robot Kingdom,* a variety of economic and business factors sparked the initial interest: a need for traditional Japanese assembly lines to become more flexible, a labor shortage, and a corporate attitude that encouraged long-term development of this new technology. By 1988, reports the U.S. National Research Council, Japan had 176,000 industrial robots, five times as many as the United States (where the industrial robot was invented!) and exceeding the entire robot population of the rest of the world.

Japan still dominates, with just less than half the world's population of robots. And while other nations are catching up, the dynamic Japanese style of robotics research has continued, with incentives from, for example, the government-funded Humanoid Robotics Project. Running from 1998 to 2002, with a budget of $38 million, the project combined government and corporate resources to develop a humanoid robot for tasks such as industrial plant maintenance, patient care, and operating construction machinery. The result, as I saw at ROBODEX 2003, is HRP-2, a 1.5 m (5 ft) tall, blue and silver, walking robot that can recover from a fall by standing up again, on its own—a feat matched only by the Sony QRIO unit.

Things are done differently in the United States. Rather than assemble massive focused programs, the government funds research on robots (and every other kind of science and technology) from a variety of sources. Some supporting agencies, such as NASA and the National Science Foundation, have civilian orientations. The research they fund is part of a climate where science and technology are meant to enhance society in general. However, a large fraction of U.S. robotics research has a different goal. That is the work in robotics supported by the Department of Defense (DoD), mostly through DARPA, which "pursues research and technology where risk and payoff are both very high and where success may provide dramatic advances for traditional military roles and missions."

Developing robots and related technologies for warfare is a worthy goal if it makes the battlefield less dangerous for humans: If we must fight wars, let us fight them with machines, not people (although the morality of this stance could be compromised if the machines are self-aware, or if the use of robot soldiers encourages some nations to believe they can go to war without any human risk or cost). And apart from its military orientation, DARPA funding has led to important results that have had some direct and positive effects on society: The Internet, for example, began under DARPA auspices. Nevertheless, there is an essential difference between targeting research for military use that might have beneficial spinoffs but might also be kept secret, and specifically aiming for civilian applications, as the Japanese do, and openly disseminating the results. Rodney Brooks, inventor of Cog, notes that when he became director of the MIT Artificial Intelligence Laboratory in the late 1990s, 95 percent of the Laboratory's research was funded by DoD. He thought that was "too much, from any perspective," and with additional corporate sponsorship, reduced the figure to 65 percent.

The differences between U.S. and Japanese research reflect national priorities and necessities. Japan has no equivalent to the enormous U.S. defense establishment, and its government funds research for civilian goals. Especially after the terror attacks of September 11, 2001, the United States is actively seeking methods to improve its security and military effectiveness, some of which fall within the science of artificial beings. One direction for military research is the development of autonomous or semiautonomous weapons. Other research areas involve combating terrorism with biometric technology such as face recognition. Much of the latter research had been carried out under DARPA's Information Awareness Office (IAO), directed by Admiral John M. Poindexter. Several IAO programs, including the Total Information Awareness project proposed in 2002, raised widespread alarm over issues of civil liberties and privacy. As a result, Congress eliminated IAO funding in late 2003, although part of these operations may be shifted elsewhere—a reminder that while we develop methods to combat terrorism, we must remain alert to possible misuse of these technologies.

Possible invasion of privacy, or worse, using the same technology that gives us wondrous robots is one dark shadow that accompanies the introduction of artificial beings into our society. Another is their potential to replace human workers, first hinted at by Aristotle when he wrote of automated machinery, and now becoming a definite possibility. According to *World Robotics 2002,* the cost of robots is falling while the cost of labor is rising. This combination presents an economic imperative that rightfully concerns the working pool, especially older workers.

Bionic technology raises a different set of concerns. There is no question about the rightness of artificial implants for the ill and injured, but what if the technology becomes so good that perfectly healthy people can augment their abilities or their lifespans at their whim? While this possibility is far distant, we have learned something from the issues swirling around other forms of human alteration such as genetic manipulation; namely, technology that modifies people in unnatural ways or overturns old definitions of birth, life, and death raises moral and legal questions, and the earlier we consider these, the better.

The alteration of people by artificial implants shares some issues with the biological modification of people, but also resolves some. One question is the familiar one of access: If a $100,000 implant can make one healthier or smarter, does that mean that only the rich will benefit from it? A brand-new concern comes from the mixed nature of a bionic individual. Imagine a person with so many implants that he or she is largely artificial. Especially if neural function has been modified, is this entity the same person who held, let us say, the right to vote and own property? This potential legal issue points to the need for new definitions of personhood and of being human. Yet the technology of artificiality can also resolve some troublesome situations. With workable artificial parts, the ill would no longer need to await donors of living tissue, dissolving the moral and medical issues surrounding the harvesting of human body parts. Another advantage of bionic modification is the fact that these alterations do not enter the gene pool—unlike genetic changes, the effects of which could include unforeseen long-term harm ongoing through the generations.

Important as all these factors are, they are not the only ones we project onto artificial beings. Religious or spiritual beliefs can also color our views toward synthetic beings. Some writers ascribe the positive attitude of the Japanese to Shinto, their native religion, and to Buddhism, imported to Japan from India in the sixth century. In his book *The Japanese Mind,* Robert Christopher comments that Buddhists take a different view of robots than do Christians because Buddhism "does not place man at the center of the universe and, in fact, makes no particular distinction between the animate and the inanimate." Along similar lines, Schodt's *Inside the Robot Kingdom* notes that Buddhism, and more especially Shinto, encompasses the belief that even inanimate things can be conscious. "Mountains, trees, even rocks are worshipped for their *kami*, or indwelling 'spirit,'" he writes, and adds,

> samurai swords and carpenter's tools have "souls"...[For a] videotape on children's robot shows, a producer writes that "people not only make friends with each other, but with animals and plants, the wind, rain, mountains, rivers, the sun and the moon. A doll [robot] in the shape of a human is therefore even more of a friend."

Within that Japanese tradition, even a Type I robot might mean more than a piece of machinery does to non-Japanese people. Some observers take this further and say that Western religion is hostile to artificial beings, the creation of which is seen as impious or worse. In science-fiction writer Stanislaw Lem's comment in Chapter 1, that an effort to make an artificial human is an attempt to "become equal to God," Lem is referring to Judeo-Christian conceptions of God. Isaac Asimov has made the same point, asserting that what he calls the "Frankenstein complex" arises in societies where God is taken as the sole creator.

But according to Anne Foerst, a theologian who has studied the religious and ethical preconceptions we bring to artificial beings, Western religious attitudes are more varied than that. Jewish belief, she writes, is "ambiguous about humanoids." On the one hand, to construct a being like the golem, as Rabbi Löw did in sixteenth-century Prague, is to praise God by exercising creativity and artisanship, which are part of God's image. On the other hand, we face the danger that humans will turn from adoring God to adoring the golem makers. The Christian tradition in the West, adds Foerst, is less ambiguous because it is more concerned with hubris, the overstepping of human bounds that angered the ancient Greek gods and remains for Christians "sin ingrained in the social consciousness."

One action that could be considered hubristic within the Western tradition, the attempt to create beings in God's image—that is, as perfect androids—might never happen, and not necessarily because it violates Christian sensibilities. The Japanese roboticist Masahiro Mori, author of *The Buddha in the Robot,* points out another reason not to attempt the construction of perfect androids. As robots like ASIMO and QRIO become more lifelike and human, the strength of our perceived connection to them rises, and feelings of threat or strangeness diminish. However, as robots become nearly identical to humans, but in some subtle way not quite so, we feel a sense of wrongness that Mori calls the "Uncanny Valley," which he advises roboticists to avoid as they design their beings.

Consider, too, the practical question: What is the value of artificial beings that are indistinguishable from humans? A generally humanoid shape is needed to operate in a world designed for the human form, and an expressive face that people can read facilitates communication, but there are not many applications where absolute fidelity to human actions and appearance is essential—except possibly in the entertainment industry, which might turn out to be a surprisingly important application, and perhaps for illegitimate uses such as those of the murderous androids in the *Terminator* films. For both psychological and pragmatic reasons, we may well find ourselves dealing, and comfortably so, with beings that look human enough rather than completely human.

In considering deadly androids and other such creatures, we might think the virtual history of artificial beings has shown us the greatest evils they could be imagined to do, but as we get closer to being able to produce highly capable beings, new and fearful possibilities arise. The poor unguided Being in Mary Shelley's *Frankenstein* suddenly seems even less monstrous; he is infinitely less threatening than a semi-autonomous military tank, say, that can recognize targets and fire on them—a possibility that Larry Matthies at NASA's Jet Propulsion Laboratory, who has worked on military robotics applications as well as planetary rovers, thinks may become a reality within 20 years. But the approaching reality can also draw on the best that creative writers have given us: the beautiful dancing cyborg Deirdre, the androids Yod and Roy Batty struggling with existential truths, the lovable robot Robbie and intelligent machine minds of *I, Robot*, and the naively charming Commander Data with his sterling qualities of honesty and loyalty.

Like any parents, we can only hope to influence our children so that they grow up both to fulfill themselves and to contribute to the world, by giving them the best start we can. Our digital children will make valuable contributions only if individual researchers, corporations, governments, and entire cultures make wise and moral choices about their purposes and uses.

HUBRIS AND HUMILITY

If we are not sure how our synthetic children will turn out, why should we embrace the difficulties of creating and nurturing them at all? One answer is that regardless of the outcome, the very act of making digital people helps us form a clearer image of what we really are as humans. Better scientific understanding of our bodies and minds is necessary if we are ever to bring artificial beings to their ultimate possibilities, but it cuts both ways because the methods used to make and study them also illuminate us. As we contemplate, and perhaps cross, the border between inert and unconscious on the one hand and living and conscious on the other—whether approached from the human or the artificial end of the spectrum—perhaps we can also throw light on the human spirit, which some call the soul. And as the theologian Anne Foerst comments, thinking about artificial personhood also makes us consider why we allow certain people into our communities and reject others—perhaps engendering a more inclusive acceptance across boundaries of race, religion, gender, and functionality as well as artificiality.

The most important benefit, however, might be a spiritual realization about our place in the universe. The specter of excessive human pride has reared its head more than once in the history of artificial beings, both virtual and real. It is an arrogance that is easy to come by in our scientific age, but not for the very greatest scientists, those whose wisdom encompasses a sense of wonder and humility as they strive to understand nature.

The great Spanish neuroanatomist Santiago Ramón y Cajal, whose work a hundred years ago laid the foundation for understanding the very brain we now struggle to emulate, felt that sense of awe. In 1906, Ramón y Cajal won the Nobel Prize in physiology for his research on the retina. Working with a staining technique developed by Camillo Golgi (who shared the prize with him), he showed for the first time separate neurons within the retina and their delicate interconnecting filaments. The retina is an outgrowth of the brain, and so this research gave us our modern picture of the nervous system and the brain as made up of separate but intricately interlinked units.

As his work and personal writings show, Ramón y Cajal was a true laboratory scientist whose first priority was the reality of facts established through painstaking effort: He might have been perfectly at home as a tough-minded member of a contemporary research team seeking to understand organic brains or make artificial ones. But despite his no-nonsense approach, what he saw in the retina lifted him to another plane and filled him with wonder. As he writes in his autobiography, he was

> amazed and confounded by the supreme constructive ingenuity revealed not only in the retina . . . but even in the meanest insect eye. There, in fine, I felt more profoundly than in any other subject of study the shuddering sensation of the unfathomable mystery of life.

Today, a century later, any person who works to artificially match or surpass what humanity is, or merely observes the effort, as I have, can only feel hubris fall away, to be replaced with awe at the complexity of what nature has wrought, humility at the difficulty of emulating it, and wonderment that we humans can yet hope to complete this astonishing journey.

Suggested Reading

In addition to books and articles in the scientific and medical literature that provided the scientific basis for *Digital People*, the popular books and articles listed below are good sources for interested readers who want more information in accessible form.

Asimov, Isaac. *I, Robot*. 1950. New York: Bantam, 1991.

Aurich, Rolf, Wolfgang Jacobsen, and Gabriele Jatho, eds. *Artificial Humans: Manic Machines, Controlled Bodies*. Berlin: Jovis, 2000.

Austen, Ian. "Learning to Speak Their Minds." *New York Times* 19 July 2002, sec. D, p. 1.

Baum, Joan. *The Calculating Passion of Ada Byron*. Hamden, CT: Archon, 1986.

Baxter, John. *Science Fiction in the Cinema*. London: Tantivy Press, 1974.

Behar, Michael. "The New Mobile Infantry." *Wired* 10.05, May 2002.

Beresford, David. "My Life as a Cyborg." *The Guardian* 3 December, 2002.

Blakeslee, Sandra. "Brain Signals Shown to Move A Robot's Arm." *New York Times* 16 November 2000, sec. A, p. 18.

________________ "In Pioneering Duke Study, Monkey Think, Robot Do." *New York Times* 13 October 2003 sec. A, p. 15.

Broad, William J. "Soon, Three New Travelers to Mars." *New York Times* 27 May 2003, sec. D, pp. 1, 4.

Brooks, Rodney A. *Flesh and Machines: How Robots Will Change Us.* New York: Pantheon, 2002.

Brumbaugh, Robert S. *Ancient Greek Gadgets and Machines.* New York: Thomas Crowell, 1966.

Bukatman, Scott. *Blade Runner*. London: British Film Institute, 1997.

Capek, Karel. *R.U.R. (Rossum's Universal Robots)*. Translated by Paul Selver and Samuel Playfair. New York: Samuel French, 1923.

Christopher, Robert. *The Japanese Mind: the Goliath Explained.* New York: Simon and Schuster, 1983.

Crick, Francis. *The Astonishing Hypothesis: The Scientific Search for the Soul.* New York: Charles Scribner's Sons, 1994.

Damasio, Antonio R. *Descartes' Error: Emotion, Reason and the Human Brain.* New York: HarperCollins, 1994.

De Camp, L. Sprague. *The Ancient Engineers.* 1960. New York: Ballantine, 1991.

Dennett, Daniel C. *Consciousness Explained.* Boston: Little, Brown and Company, 1991.

Dick, Philip K. *Blade Runner: Do Androids Dream of Electric Sheep?* 1968. New York: Ballantine, 1982.

Dreifus, Claudia. "Do Androids Dream? M. I. T. Is Working on It." *New York Times* 7 November 2000, sec. F, p. 3.

_____________ "A Passion to Build a Better Robot, One With Social Skills and a Smile." *New York Times* 10 June 2003, sec. D, p. 3.

Edelman, Gerald M., and Giulio Tononi. *A Universe of Consciousness: How Matter Becomes Imagination.* New York: Basic Books, 2000.

Eisenberg, Ann. "Blind People With Eye Damage May Someday Use Chips to See." *New York Times* 24 June 1999, sec. E, p. 7.

__________ "Restoring the Human Touch to Remote-Controlled Surgery." *New York Times* 30 May 2002, sec. E, p. 7.

__________ "Teaching Machines to Hear Your Prose and Your Pain." *New York Times,* 1 August 2002, sec. E, p. 7.

Elsaesser, Thomas. *Metropolis.* London: British Film Institute, 2000.

Fass, Allison. "Speak Easy." *Forbes,* 6 January 2003, pp. 135-136.

Fillon, Mike. "The New Bionic Man." *Popular Mechanics,* February1999, pp. 51-55.

Ford, Kenneth M., Clark Glymour, and Patrick J. Hayes, eds. *Android Epistemology.* Cambridge, MA: MIT Press, 1995.

Foreman, Judy. "The 43 Facial Muscles That Reveal Even the Most Fleeting Emotions." *New York Times* 5 August 2003, sec. D, pp. 5, 8.

Freud, Sigmund. "The Uncanny." *The Standard Edition of the Complete Psychological Works of Sigmund Freud,* Vol. XVII. James Strachey, ed. London: Hogarth, 1955. pp. 219-252.

Gardner, Howard. *Intelligence Reframed.* New York: Basic Books, 1999.

Gorman, James. "Fishing for Clarity in the Waters of Consciousness." *New York Times* 15 May 2003, sec. D, p. 3.

Graham-Rowe, Duncan. "Second Sight." *New Scientist* 23 November 2002, pp. 34-37.

Gray, Chris Hables. *The Cyborg Handbook.* New York: Routledge, 1995.

Greenberg, Ilan. "A Nose for Business." *Technology Review* July/August 1999, pp. 63-67.

Haining, Peter, ed. *The Frankenstein Omnibus.* Edison, NJ: Chartwell Books, 1994.

Harnad, Stevan. "No Easy Way Out," *The Sciences* Spring 2001, pp. 36-42.

Hodges, Andrew. *Alan Turing: The Enigma.* New York: Walker & Company, 2000.

Hogan, James P. *Mind Matters: Exploring the World of Artificial Intelligence.* New York: Ballantine, 1997.

James, Peter, and Nick Thorpe. *Ancient Inventions*. New York: Ballantine, 1994.

Kirsner, Scott, "Making Robots, with Dreams of Henry Ford," *New York Times* 26 December 2002, sec. E, pp. 1, 3.

Kurzweil, Ray. *The Age of Intelligent Machines.* Cambridge, MA: MIT Press, 1990.

___________ *The Age of Spiritual Machines: When Computers Exceed Human Intelligence.* New York: Penguin Books, 1999.

___________ and Mitchell Kapor. "Yes or No: A Computer Will Pass the Turing Test by 2029." *Discover* May 2002, p. 123.

Lakoff, George, and Mark Johnson. *Philosophy in the Flesh: The Embodied Mind and its Challenge to Western Philosophy*. New York: Basic Books, 1999.

Lang, Fritz, Paul Jensen, and Siegfried Kracauer. *Metropolis.* 1973. London: Faber and Faber, 1989.

"Lord of the Robots: Q&A with Rodney Brooks." *Technology Review* April 2002, pp. 80-82.

Malone, Robert. *The Robot Book.* New York: Push Pin Press, 1978.

Maloney, Mack. *Planet America.* New York: Ace, 2001.

Markoff, John. "Computing's Big Shift: Flexibility in the Chips." *New York Times* 16 June 2003, sec. C, pp.1, 4.

Marquis, Christopher. "The Right and Wrong Stuff of Thinking Outside the Box." *New York Times* 31 July 2003, sec. A, p. 15.

Matusomoto, Chie. "Powerful Astro Boy Makes Colorful Comeback on the Little Screen." *The Asahi Shimbun* 5-6 April 2003, p. 32.

Maxford, Howard. *The A-Z of Science Fiction & Fantasy Films*. London: BT Batsford, 1997.

Mazlish, Bruce. "The Man-Machine and Artificial Intelligence." *Stanford Electronic Humanities Review* 4, No. 2.

McNeil, Daniel. *The Face.* Boston: Little, Brown and Company, 1998.

Menzel, Peter and Faith D'Aluisio. *Robo Sapiens: Evolution of a New Species.* Cambridge, MA: MIT Press, 2000.

Minsky, Marvin. *The Society of Mind.* New York: Simon and Schuster, 1986.

Moore, C. L. "No Woman Born." *A Treasury of Science Fiction*. Groff Conklin, ed. New York: Crown, 1948. pp. 164-201.

Moravec, Hans. *Robot: Mere Machine to Transcendent Mind.* New York: Oxford University Press, 1999.

Mori, Masahiro. *The Buddha in the Robot.* Translated by Charles S. Terry. Tokyo: Kosei, 1981.

Napier, John. *Hands* (Revised by Russell H. Tuttle). 1980. Princeton, NJ: Princeton University Press, 1993.

Nicolelis, Miguel A.L. and John K. Chapin. "Controlling Robots With the Mind." *Scientific American* October 2002, pp. 46-53.

Penrose, Roger. *Shadows of the Mind: A Search for the Missing Science of Consciousness.* Oxford: Oxford University Press, 1996.

_____________ *The Emperor's New Mind: Concerning Computers, Minds, and the Laws of Physics.* Oxford: Oxford University Press, 2002.

Pera, Marcello. *The Ambiguous Frog.* Translated by Jonathan Mandelbaum. Princeton, NJ: Princeton University Press, 1992.

Perkowitz, Sidney. "In Salmon do did Mobile Bond," *New Scientist* 19 and 26 December 1998 and 2 January 1999, pp. 62, 63.

_____________ "Feeling is Believing," *New Scientist* 11 September 1999, pp. 34-37.

Picard, Rosalind. *Affective Computing.* Cambridge, MA: MIT Press, 1997.

Piercy, Marge. *He, She and It.* New York: Alfred A Knopf, 1991.

Plotz, David. "I Spy With My Eagle Eye." *Slate,* 5 March 2003.

Quain, John. R. "It Mulches, Too? Robotic Mowers Win a Following." *New York Times* 31 July 2003, sec. E, pp. 4, 8.

Ratlif, Evan. "Born to Run." *Wired,* 9.07, July 2001.

Regis, Ed. *Great Mambo Chicken and the Transhuman Condition.* Reading, MA: Addison-Wesley, 1990.

Rosenfeld, Israel. *The Strange, Familiar, and Forgotten: An Anatomy of Consciousness.* New York: Alfred A. Knopf, 1992.

Sacks, Oliver. "To See or Not to See." *The New Yorker* 10 May 1993, pp. 59-73.

Schodt, Frederik. *Inside the Robot Kingdom: Japan, Mechatronics and the Coming Robotopia.* Tokyo: Kodansha, 1988.

Searle, John B. *The Mystery of Consciousness.* New York: New York Review of Books, 1997.

Segel, Harold B. *Pinocchio's Progeny: Marionettes, Automatons, and Robots in Modernist and Avant-Garde Drama.* Baltimore: Johns Hopkins University Press, 1995.

Senn, Bryan and John Johnson, eds. *Fantastic Cinema Subject Guide.* Jefferson, NC: McFarland, 1992.

Shenon, Philip "New Devices to Recognize Body Features on U S. Entry." *New York Times* 30 April 2003, sec. A, p. 16.

Shelley, Mary. *Frankenstein.* 1818. Introduction, Jeffery Deaver. Oxford: Oxford University Press, 1969, 2001.

_____________ *Frankenstein.* 1831. Introduction, Diane Johnson. New York: Bantam, 1991.

Sundman, John. "Artificial Stupidity." *Salon,* Part 1, 26 February 2003; Part 2, 27 February 2003.

Swade, D. "Redeeming Charles Babbage's Mechanical Computer." *Scientific American* February 1993, pp. 86-91.

Tellotte, J. P. *Replications: A Robotic History of the Science Fiction Film.* Urbana, IL: University of Illinois Press, 1995.

Templado, Louis. "Cosmic Ranger, Laugh at Danger—Happy Birthday, Astro Boy!" *The Asahi Shimbun* 5-6 April 2003, p. 32.

"The Talking and Listing Washing Machine." *Popular Science,* June 2003, p. 26.

Toner, Mike. "The Road to Tera." *The Atlanta Journal-Constitution* 20 July 2003, sec. C, pp. 1, 6.

Turing, A.M. "Computing Machinery and Intelligence." *Mind* **59**, 433-460 (1950).

Usher, Albert Payson. *A History of Mechanical Inventions.* 1954. New York: Dover, 1988.
Vance, Ashlee. "Robotic Road Trip on a Military Mission." *New York Times* 9 October 2003, sec. E, pp. 1, 6.
Wayner, Peter. "As Plain as the 'Nose' on Your Chip." *New York Times* 8 July 1999, sec. D, p. 11.
Wegner, Daniel. *The Illusion of Conscious Will.* Cambridge, MA: MIT Press, 2002.
Whynott, Douglas. "The Robot That Loves People." *Discover* October, 1999, pp. 66-73.
Williamson, Jack. "With Folded Hands." *A Treasury of Science Fiction.* Groff Conklin, ed. New York: Crown, 1958. pp. 129-164.
Wood, Gaby. *Edison's Eve.* New York: Alfred A. Knopf, 2002.

Filmography

A. I.: Artificial Intelligence (2001)
Blade Runner (1982, director's cut 1992)
Bride of Frankenstein (1935)
The Day the Earth Stood Still (1951)
Firefox (1982)
Forbidden Planet (1956)
Frankenstein (1910)
Frankenstein (1931)
The Ghost of Frankenstein (1942)
Jason and the Argonauts (1963)
Metropolis (1927)
One Flew Over the Cuckoo's Nest (1975)
RoboCop (1987)
The Six Million Dollar Man (1973–1978)*
Star Trek: The Next Generation (1987–1994)*
Star Trek: Generations (1994).
Star Wars (1977).
Star Wars: Episode V—The Empire Strikes Back (1980)

---

*Television series

The Terminator (1984)
Terminator 2: Judgment Day (1991)
Terminator 3: Rise of the Machines (2003)
2001: A Space Odyssey (1968)
The Wizard of Oz (1939)

Acknowledgments

I am grateful to my dear wife Sandy for living well with my physical and mental absence while I wrote *Digital People*. Her acceptance, and her role as a patient and insightful sounding board and editor, helped enormously. My son Mike, trained in AI, was a valuable consultant. He also showed me the pop culture of robots through his band *R.U.R.* and its great T-shirt; and by tracking down the song "I am Electro" by Meat Beat Manifesto, inspired by the popular 1939 New Work World's Fair robot Elektro.

The people and resources of Emory University make it a wonderful place for a writer. Robyn Fivush, Scott Lilienfeld, visitor Jeffrey Mullins, Darryl Neill, Leslie Taylor, and Elaine Walker gave generous help by reading the manuscript or lending their expertise. The Consciousness Seminar, led by Arri Eisen and Howard Kushner, provided useful background. Kate Bennett, David Schaar, Michael Gadbaw, and Lauren Gunderson helped with research.

Special thanks go to Philip Hammer and Charles Penniman of Philadelphia's Franklin Institute for demonstrating the "Draughtsman-Writer" automaton, and to Jim Randolph for sharing his encyclopedic knowledge of artificial beings in the media. Terry Gips and Isa Gordon helped me think about the book at its beginning, as did Phil

Schewe, who also read the manuscript. Others who helped include Peter Brown, Amy Bruckman, Kenji Ito, Massimo Piccardi, Jeffrey Reznick, Jeffrey Shoap, Graziella Tonfoni, and Eric Willadsen. My deep thanks also to the researchers at Carnegie Mellon University, MIT, Georgia Tech, JPL, the Sony Corporation and elsewhere, who gave me generous amounts of time in person or by telephone and e-mail.

My agent Michelle Tessler of Carlisle & Company has been steadfastly enthusiastic about the pop culture of robots. At Joseph Henry Press, Jeffrey Robbins provided dedicated and incisive editing that greatly improved the book, and Maire Murphy provided helpful copyediting.

None of these people is responsible in the least for any errors or misstatements that might appear in the book.

Having delved into the world of artificial beings, I'm ready to rejoin family and friends in the human world; but it won't be long before we all number synthetic creatures among our friends.

Sidney Perkowitz
Atlanta, GA

Index

A

A. I.: Artificial Intelligence (2001), 9, 17, 47–48, 209, 225
Abacus, 64
Adam, 20
Adaptive computer chips, 188, 207
Adaptive optics, 172
Adolphs, Ralph, 162
Advanced step in innovative mobility. *See* ASIMO the walking robot
Affective Computing, 182–183
The Age of Spiritual Machines, 201
AIBO robot dog, 3, 169, 181, 221
Akoulathon, 97
Aldini, Giovanni, 22
Aldiss, Brian, 47
Ambrose, Robert, 165
Amtrak call system, 163–164
Amyotrophic lateral sclerosis (ALS), 192
Analytical Engines, 65
Androids, 1–13, 18
  defined, 4
  limited abilities of, 10
"Animal electricity," 22, 62
Aquinas, Thomas, 20
Argonauts, 19
Ariel, 37
Aristotle, 5, 62, 214
Arithmetic-logic unit (ALU), 176
Arkin, Ronald, 11, 174, 181
Artificial beings
  being bionic, 85–102
  built-in inheritance of, 49
  creation of, 7, 24
  going out of control, 20
  longing to join the human race, 24
  meaning and history, 15–102
  real history of, 51–84
  self-awareness in, 21
  virtual history of, 17–50
Artificial intelligence (AI), 13, 38, 71
  advances in, 10
Artificial Muscle Research Institute, 143
Artificial smell, 166–167
Artificial speech, 162
Artificial taste, 167–168
Artificial touch, 165

ASIMO the walking robot, 52, 130–131, 134, 144, 176–177, 196, 203, 216
Asimov, Isaac, 9, 18, 31, 49, 72, 184, 216
The Astonishing Hypothesis, 113
Astro Boy, 36, 212
Atomic bomb, 31, 37
Automata, 56–61
Automated workers, 5
The Automaton Theater, 53–54

B

Babbage, Charles, 60, 65–67, 69
Bacon, Roger, 20
Bar-Cohen, Yoseph, 142–144
Barnum, P.T., 58
Baum, L. Frank, 25
Behavior-Based Robotics, 11, 174
Bell, Alexander Graham, 78
Bell Labs, 68, 80
Binary numbers, 68, 70
Binet, Alfred, 174
Biomechatronic hand, 137–138
Biomedicine, advances in, 11
Biometrics, 155
"Biomorphic" chips, 171
Bionic artificial beings, 85–102
  charging the body, 92–94
  digital ears, 97–101
  electrifying the mind, 94–97
  with feeling, 101–102
  unfeeling limbs, 87–92
Bionic humans, 3–5, 18
BioRobotics Laboratory, 166
Biotechnology, 38
Birnbaum, Lawrence, 205
Blade Runner (1982, director's cut 1992), 9, 18, 39–41, 225
Blind, restoring sight to, 170–171
Boahen, Kwabena, 171, 187
Bolshevik worker's revolution, 26
Boole, George, 67
Boolean algebra, 67
Brain, role of
  in artificial beings, 174
  in humans, 174
Brain implants, 170–171, 184–189, 210–211, 214
Brain-machine interfaces (BMI), 189–193
Breazeal, Cynthia, 2, 142, 144, 177–178, 181–182, 196, 208
Bride of Frankenstein (1935), 29, 225
Brooks, Rodney, 1–2, 75–76, 120, 127, 178, 182, 208
The Buddha in the Robot, 216
Buddhist outlook, 215
Byron, Lord, 21, 65

C

Caesar, Julius, 7
Caidin, Martin, 39
Calculators, mechanical, 64–65
Calculus, 64
Calvin, Susan, 31–32, 49
Capek, Karel, 9, 18, 25, 72, 140, 201, 210
Carbon arc light, 62
Carnegie Mellon University, 2
Chalmers, David, 108
Chapin, John, 137–138
Charging the body, 92–94
Chess playing, computational technology used for, 60
Chinese Room scenario, 117
Chips, getting smarter, 201–203
Chow, Vincent and Alan, 170
Christian tradition, 215–216
Christopher, Robert, 215
Clarke, Arthur C., 38
Clocks and clockmakers, 54–55, 58
Cochlear implantation, 98–99, 102
Cog robot, 1–2, 208, 213
Coiled springs, 54
Companion race, creating, 3, 12
Computational technology

advances in, 10, 38, 73–74
affective, 182–183
used for chess playing, 60
"Computing Machinery and Intelligence," 208
"Conditional jumps," 66, 68–69
Consciousness
criteria for, 196, 210
higher-order, 194–196
"overrated," 11
signs of in living machines, 195–197
Consciousness Explained, 112
Coppélia, 25
Crawling, 131–133
Creation of artificial beings, 7, 24
from bronze and clay, 18–21
Crick, Francis, 108, 113
Cyborglike activity, 190
Cyborgs, 18
defined, 5
Cyrano Sciences, 167

D

Da Vinci system, 165
DaimlerChrysler Corporation, 155
Damasio, Antonio, 110, 183
Dante II, 127
Dario, Paolo, 134, 138
Data, Lieutenant Commander, 9, 46–47, 158, 184, 211, 217
Davy, Humphry, 62
The Day the Earth Stood Still (1951), 37, 128, 225
De Juan, Eugene, 170
De Vaucanson, Jacques, 56–58
Dead bodies, 7
"Deep Blue" computer, 60, 74
Defense Advanced Research Projects Agency (DARPA), 8, 133, 193
communicator project, 8, 133, 213
information Awareness Office (IAO), 214
Deirdre, 18, 29–30, 49, 85, 190, 195, 217
Delibes, 25
Dennett, Daniel, 112, 116
Department of Defense, 8, 133, 154, 200–201
Descartes, René, 55, 107, 112, 159
Desoutter, Marcel, 90
Devol, George, 73
Dick, Philip K., 39
Difference Engines, 65
Differential Analyzer, 67–68
Digital creatures
hearing, 97–101
thinking, 116–119
Digital Creatures Laboratory, 180–181
Digital electronics, 13, 20
advances in, 10, 63
Discourse on Method, 55–56
DLR Hand II, 165
DNA alphabet, 20, 108
Do Androids Dream of Electric Sheep?, 39
Dobelle, William, 171
Dolls, coming to life, 7
"Draughtsman-Writer," 58–60
"Dream Replicants of the Cinema," 32
Drug delivery systems, implanted, 4

E

Earth Simulator computer, 185
Eckert, J. Presper, 69
Edelman, Gerald, 113–114, 195, 207
Edison, Thomas, 78
Ekman, Paul, 158
Electrical batteries, 54
Electricity
practical use of, 61–63
static, 61
Electrifying the mind, 94–97
Electro-active polymers (EAP), 142, 144
Electroconvulsive therapy (ECT), 94–95

Electroencephalography (EEG), 109
Elektro (robot), 30, 72, 81
Emory Univeristy, 192
Emotions, link with reason, 183–184
Endoscopy, 165–166
Engelberger, Joseph, 73
ENIAC computer, 69
Enigma code breaking, 69–70
Enlightenment era, 56
  "smile of," 58
Enterprise, Starship, 46
"Evidence," 34
Exoskeletons, activating, 193

F

Facial Action Coding System, 158
Facial identity, 155–159
Faraday, Michael, 63
Female robots, bad and beautiful, 27–30
Firefox (1982), 193, 225
Five senses, and beyond in living machines, 147–172
  extending human abilities, 168–169
  extending human senses, 170–172
  facial identity, 155–159
  moving from one location to another, 152–155
  relating to others, 159–164
  synthetic vision, 149–152
  touching an android, 164–168
Foerst, Anne, 215, 218
Forbidden Planet (1956), 37, 225
Frankenstein: The Modern Prometheus (1818), 7, 9, 17, 21–25, 208–209, 217
Frankenstein, Victor, 7, 18, 21–25, 49, 189
Frankenstein (1910), 23, 225
Frankenstein (1931), 21, 23, 225
Frankenstein's creature, or Commander Data, 4, 199–219
Franklin, Benjamin, 61
Franklin Institute, 58
Fraunhofer Institute for Integrated Circuits, 156
Frederson, John, 27–28
Freud, Sigmund, 7
Friesen, Wallance, 158
Fromherz, Peter, 194–195
Fuita, Masahiro, 180
Functional magnetic resonance imaging (fMRI), 109–110

G

Galvani, Luigi, 22, 62–63
Galvanism, 22
Gardner, Howard, 175–176, 196–197, 202, 207
Genghis robot, 75–76
Georgia Institute of Technology, 11
The Ghost of Frankenstein (1942), 225
Gödel's theorem, 115
Godwin, Mary Wollstonecraft. *See* Shelley, Mary
Goethe, Johann Wolfgang von, 57
Golems, 20
Golgi, Camillo, 218
Gort the robot, 37, 48, 128
"Gray matter," 184–185
Greek myth, 9, 18–19, 52–53, 87

H

Hal 9000, 38–39, 209
Hands, 134
Hanson, David, 143
Haptic senses, 147, 164, 166
Harbou, Thea von, 27
Hawking, Stephen, 65, 192
He, She and It (1991), 39, 44–46, 49, 209, 211
Henlein, Peter, 55
Hephaestus, 19, 84, 87
Heron of Alexandria, 53–54
Higher-order consciousness, 194–196

Hirose, Shigeo, 141
Hirschberg, Julia, 160
History of artificial beings, 15–102
  real, 51–84
  virtual, 17–50
Hoffman, E.T.A., 9, 25
Hogan, James, 204, 208
Honda Corporation, 52, 124, 128, 130, 176
Honda research laboratories, 2
Howe, Robert, 166
Hubris, 6, 8, 24, 216, 218–219
Human abilities, extending, 168–169
Human race, artificial beings longing to join, 24
Human senses, extending, 170–172
Humanity, meaning of, 39, 211
Humanoid Robotics Project, 212
The Humanoids, 34
Humayun, Mark, 170

I

I, Robot, 9, 31–34, 49, 72, 184, 217
I-Cybie robot dog, 3, 160
The Illusion of Conscious Will, 112
Implant science, 38
Implanted drug delivery systems, 4
Indian mythology, 6
Information Awareness Office (IAO), 214
Infrared vision, 168–169
Inheritance of artificial beings, built-in, 49
Intuitive Surgical, 165
Inside the Robot Kingdom, 212, 215
Intelligence
  defining, 174–176
  embodied in living machines, 119 122
International Space Station, 135
Internet, 169, 173, 213
Interpersonal intelligence, 177
Investigation into the Laws of Thought, 67
IQ testing, 175
iRobot Co., 126

J

Jacquard loom, 58, 69
James, William, 105–106
The Japanese Mind, 215
Jaquet-Droz, Pierre, 56, 58
Jason and the Argonauts (1963), 19, 225
Jews of Prague, 20
Johnson, Mark, 120
Johnson Space Flight Center, 135
Judeo-Christian tradition, 216

K

Kanade, Takeo, 156, 158
Karloff, Boris, 21
Kasparov, Gary, 60
Kempelen, Wolfgang von, 60
Kennedy, Philip, 191–193, 210–211
Kinesthetic sensing, 164–165, 180
King, Augusta Ada, 65
Kismet robot, 2, 12, 142, 178–179, 208
  showing feelings, 181–182, 196–197
Klaatu, 37
Koch, Christopher, 113
Kubrick, Stanley, 38
Kurzweil, Ray, 201–202, 205

L

Lakoff, George, 120
Land mines, 91
Lang, Fritz, 18, 27
Language use by, 203–206
Laser surgery, 172
Learning, 207–211
Lem, Stanislaw, 6, 216
Lemonnier, Anicet-Charles-Gabriel, 58
L'Homme Machine, 56
Liang, Junzhong, 172

Life electric, 61–64
Limbs, movement, and expression in living machines, 123–145
  arms, hands, fingers, and thumbs, 133–140
  crawling and morphing, 131–133
  eight legs to two, 127–131
  smoothing the motion, 140–145
  wheeled, treaded, and tracked, 125–127
Littlewort, Gwen, 158
Living machines, 103–219
  the five senses, and beyond, 147–172
  frankenstein's creature, or Commander Data, 4, 199–219
  limbs, movement, and expression, 123–145
  mind-body problems, 105–122
  thinking, emotion, and self-awareness, 173–197
Loebner, Hugh, 204
Loebner Prize, 204
Looking human, 81–84
Löw, Rabbi, 20, 216

M

Machine morality, 18, 31–38
MachineVision group, 150
Magnus, Albertus, 23, 159
Maillardet, Henri, 58–60, 81
Maloney, Mack, 19
"The Man-Machine and Artificial Intelligence," 56
Manhattan Project, 31
"Mark I" computer, 68–69
Master, Paul, 188
Materials science, advances in, 10
Matthies, Larry, 150, 153, 217
Mauchly, John W., 69
Max Planck Institute for Biochemistry, 194
Mazlish, Bruce, 56
McCarthy, John, 71
McCulloch, Warren, 78
Mead, Carver, 187
Meaning of artificial beings, 15–102
Mechanics, 53
Mechatronic enginering, 125
"Mechatronics," 52
Metropolis (1927), 18, 25, 39, 56, 225
Mettrie, Julien Offroy de La, 56, 58
Mind-body problems of living machines, 105–122
  digital creatures thinking, 116–119
  dueling theories, 111–115
  embodied intelligences, 119–122
  no easy answer, 106–110
Mind Matters: Exploring the World of Artificial Intelligence, 204
Minos, King, 19
Minsky, Marvin, 71, 76–77, 206, 209
MIT, 1–2, 71
  artificial Intelligence Laboratory, 213
  media Lab, 182
Molecular biology, advances in, 10
Moleschott, Jakob, 56
Moore, C.L. (Catherine Lucille), 18, 29, 190
Moral choices, 211–217
Moravec, Hans, 118, 182–183, 186, 196, 201–202
Mori, Masahiro, 216
Morphing, 131–133
Motion
  bodies in, 52–61
  from one location to another, 152–155
  smoothing, 140–145
Mount Olympus, 19
Moxon, Karen, 138
Mozziconacci, Sylvie, 161
Murphy, Alex, 39, 44
Mussa-Ivaldi, Sandro, 190
My Real Baby toy robot infant, 3
The Mystery of Consciousness, 113

N

Nanocyborgs, 194
Nanotechnology, advances in, 10
Napier, John, 134–135
Napoleon, 60, 62
NASA robotic planetary explorers, 13, 152–153
National Aeronautics and Space Administration (NASA), 39, 135, 152, 165, 203, 217
National Research Council, 212
Neural activity, magnetic fields produced by, 193–194
Neural Signals, 192
Neuromorphic approach, 187
"Neuromorphic" chips, 171, 207
"Neuron on a chip," 194
"Neurorobotics," 138
New York World's Fair (1939), 30, 72
Nicolclis, Miguel, 191, 193
"No Woman Born," 18, 29, 190
Notation, positional, 64
Nursebot, 123
Nutcracker, 25

O

Offenbach, Jacques, 25
Olympia, 25
One Flew Over the Cuckoo's Nest (1975), 95, 225
Opportunity robot, 152–153
Optobionics Corporation, 170
Ozma of Oz, 25

P

PackBots, 126–127
Palo Alto Research Center, 2
Pappus of Alexandria, 53
Parallel processing, 207
Paranoia factor, 169
Parkinson, James, 96
Pathfinder, 135
Penrose, Roger, 115
Pentium chips, 185–186
Philon of Byzantium, 53
Philosophy in the Flesh: The Embodied Mind and its Challenge to Western Philosophy, 120
Picard, Rosalind, 182–184
Piercy, Marge, 44–45, 209, 211
Pinocchio, 9, 47
Pirjanian, Paolo, 153
Plato, 53, 62
Pliny the Elder, 62
Poe, Edgar Allen, 60
Poindexter, John M, 214
PolyBots, 132, 140
Positional notation, 64
Positron emission tomography (PET), 109–110
Prime Directive, 35–36
Privacy issues, 214
Prostheses, 87–92
Pygmalion, 9, 18, 48

Q

QRIO robot, 177, 181, 196, 200, 203
QuickSilver Technology, 188

R

R2D2, 17
Rachael, 18, 40
Ramón y Cajal, Santiago, 218–219
Ray, Johnny, 192–193, 195, 211
Real history of artificial beings, 51–84
  bodies in motion, 52–61
  the life electric, 61–64
  looking human, 81–84
  on and off, 67–78
  sensing, 78–81
  thinking human, 64–67

Relating to others, in living machines, 159–164
Relays, 68
"Robbie" the robot, 32–33, 217
Robby the robot, 37
RoboCop (1987), 18, 39, 42–44, 85, 183, 225
ROBODEX 2003 exposition, 162, 177, 180, 212
Robonaut, 135–138, 165, 203
Robot: Mere Machine to Transcendent Mind, 118, 182–183
Robot armies, 25–27
Robot dogs, 3, 12
Robotic soccer, 169
Robots, 4, 18, 199–219, 213
  chips getting smarter, 201–203
  defined, 4
  for hazardous duty, 8
  industrial, 73, 199
  language use by, 203–206
  learning, 207–211
  limited abilities of, 10
  in living machines, 4, 199–219
  military, 201, 213
  moral choices, 211–217
  pride and humility, 24, 218–219
  self-awareness, 206–207
Roomba, 126
"Runaround," 31
R.U.R. (Rossum's Universal Robots) (1921), 10, 18, 25–27, 30, 49, 72, 140, 169, 201, 210

S

"The Sandman," 9, 25
Sarcos Corporation, 135
Schank, Roger, 205
Schodt, Frederik, 212, 215
Scott, Ridley, 39
Searle, John, 112–113, 117
Seattle, John, 110
Seeßlen, Georg, 32
Seiko Epson Corporation, 169
Self-awareness in artificial beings, 21, 206–207
Senses, 78–81
  functions of, 172
Sensors, density of, 165
The Sentinel (1951), 38
Severe acute respiratory syndrome (SARS), 169
Shadows of the Mind: A Search for the Missing Science of Consciousness, 115
Shakespeare, William, 37
"Shakey" the robot, 74–75, 79
Shannon, Claude, 67–68, 71
Shelley, Mary, 7, 17, 21–25, 65, 217
Shelley Percy Bysshe, 21–23
Shinto, 215
Shriberg, Elizabeth, 161
Silverlit Toys, 3
The Six Million Dollar Man (1973-1978)—television series, 18, 39, 42–44, 225
Smart digital beings, 176–181
Soccer, robotic, 169
The Society of Mind, 76, 206
Sony Corporation, 3, 180–181
  digital Creatures Laboratory, 180–181
Speech Experts, 162
Speech recognition systems, 160–162
Spielberg, Steven, 47
Spirit robot, 152–153
Stanford Research Institute, 74
Star Trek: Generations (1994), 225
Star Trek: The Next Generation (1987-1994)—television series, 4, 46, 158, 184, 211, 225
Star Wars: Episode V—The Empire Strikes Back (1980), 225
Star Wars (1977), 17, 225
Static electricity, 61
Sternberg, Robert, 175
STMicroelectronics Corporation, 166

Stochastic Neural-Analog Reinforcement Computer (SNARC), 78
Stolcke, Andreas, 161
Surgery, minimally invasive, 165
Survival factors, 182–183
Symbols and artificial beings, 20
Synthetic vision, 149–152

T

"The Tales of Hoffman," 25
Talos robot, 19, 48
Tchaikovsky, Piotr Ilyich, 25
The Tempest, 37
Terminator 2: Judgment Day (1991), 42, 226
Terminator 3: Rise of the Machines (2003), 226
The Terminator (1984), 18, 39, 41–42, 217, 226
Thinking, emotion, and self-awareness in living machines, 64–67, 173–197
  brain-machine interfaces, 189–195
  brains for artificial bodies, 184–189
  emotional capabilities, 181–184
  signs of consciousness, 195–197
  smart digital beings, 176–181
Thomas, Craig, 193
Three Laws of robotics, 31–32, 35, 49, 211
Thrun, Sebastian, 151
Tiger Electronics, 3
Tik-Tok, the "Machine Man," 25
Tin Woodman, 25, 85
Tononi, Giulio, 113–114
Total Information Awareness project, 214
Touching an android, 164–168
Turing, Alan, 10, 15, 70–71, 108, 164, 204, 208–209
Turing test, 10, 15, 71, 144, 159, 203–205
"The Turk," 60
Turkle, Sherry, 3
2001: A Space Odyssey (1968), 209, 226

U

"The 'Uncanny'", 7
"Uncanny Valley," 216
Une Soirée chez Madame Geoffrin, en 1755, 58
UNIMATE, 73
United Nations Economic Commission for Europe, 199
UNIVAC computer, 70
A Universe of Consciousness: How Matter Becomes Imagination, 113

V

Vacuum tube operation, 69
Vagus nerve stimulation (VNS), 95–96
Victor's creation, 21–25
Virtual faces, 157–158
Virtual history of artificial beings, 17–50
  all too human, 38–50
  creatures of bronze and clay, 18–21
  female robots—bad and beautiful, 27–30
  machine morality, 18, 31–38
  robot armies, 25–27
  victor's creation, 21–25
Visual simultaneous localization and mapping (VSLAM), 154
Volta, Alessandro, 62–63
Voltaire, 58
Von Neumann, John, 69–70
Vulcan, 19

W

Walt Disney theme parks, 2, 82, 178
Walter, Gerard, 7

War on terrorism, 162
Warwick, Kevin, 170
Watson, James, 108
Watt, James, 61
Wave-front sensors, 172
Wegner, Daniel, 112
Westinghouse Company, 30, 72
Whale, James, 21
Wilkes, Maurice, 70
Wilkins, Maurice, 108
Williams, David, 172
Williamson, Jack, 10, 34
"With Folded Hands," 10, 34, 49, 169
The Wonderful Wizard of Oz (1939), 25, 226
World RoboCup event, 169
World Robotics 2002, 199, 214
World Trade Center attack, 126
World's Columbian Exposition (1893), 61

Y

Yod, 44–46, 209–211, 217

Z

Zuse, Konrad, 68